D0881714

The Matching Law

RICHARD J. HERRNSTEIN

The Matching Law

Papers in Psychology and Economics

Edited by
HOWARD RACHLIN *and* **DAVID I. LAIBSON**

RUSSELL SAGE FOUNDATION
New York

HARVARD UNIVERSITY PRESS
Cambridge, Massachusetts & London, England   1997

Copyright © 1997 by the President and Fellows of Harvard College
All rights reserved
Printed in the United States of America

Library of Congress Cataloging-in-Publication Data
Herrnstein, Richard J.
  The matching law : papers in psychology and economics / Richard J.
Herrnstein ; edited by Howard Rachlin and David I. Laibson.
    p. cm.
  Collection of previously published articles by Richard J.
Herrnstein.
  Includes bibliographical references and index.
  ISBN 0-674-06459-3 (alk. paper)
  1. Reinforcement (Psychology) 2. Choice (Psychology)
3. Mathematical optimization. 4. Economics—Psychological aspects.
I. Rachlin, Howard, 1935–. II. Laibson, David I. III. Title.
  [DNLM: 1. Herrnstein, Richard J. 2. Reinforcement (Psychology)—
collected works. 3. Choice Behavior—collected works. BF
319.5.R4 H568m 1997]
BF319.5.R4H47   1997
153.8'3—dc20
Shared Cataloging for DNLM
Library of Congress   96-31500

Contents

SOCIAL SCIENCES DIVISION
CHICAGO PUBLIC LIBRARY
400 SOUTH STATE STREET
CHICAGO, IL 60605

Preface

Richard Herrnstein (1930–1994) is best known by the public and by the broader scientific community for his books on crime, on intelligence, and on social policy. In psychology, however, especially among psychologists studying choice behavior, and in economics, especially among economists dissatisfied with traditional optimization theory, he is better known for a general principle of choice called the matching law. In the articles selected for this volume Herrnstein presents evidence for the matching law as a general principle of choice and extends this principle, discovered in the animal laboratory, to human clinical psychology (self-control, addiction), biology (selection mechanisms, foraging), and economic choice.

At the time of his death Herrnstein was working on a new book to be called *Behavioral Economics: Toward a Science of Human Choice*. This book would have been based on the articles that appear here. They constitute Herrnstein's most important contribution to behavioral science.

Howard Rachlin
David I. Laibson
May 1996

Editors' Note

When it became clear in 1994 that Richard Herrnstein would not be able to finish the book he was then working on, we volunteered to assist him in putting together a collection of the articles on which the book was to be based. Herrnstein agreed and checked off on his vita the articles he thought might go into such a book. Chapters 3, 4, 6, 7, 8, 9, 10, 11, and 15 of the present volume were checked. In addition, Herrnstein and Vaughan (1980), Herrnstein (1988), and two papers then in manuscript form, "Dearer by the Dozen: A Pricing Scheme for Meliorizers" and "Giffen Goods in Rats: A Reply to Battalio et al." (with Dražen Prelec), were checked. After Herrnstein's death we felt it would be appropriate to widen the scope of the collection to include several earlier papers seminal for the development of the matching law. There were two stages to this development: first, Herrnstein's break with Skinner's reflex-based psychology wherein Herrnstein adopted a molar view of behavior; second, his rejection of optimization (identified by him with rationality). The articles proposed for inclusion thus far were all part of the second stage, but we wanted to include papers relevant to the first stage as well. Demands of space and considerations of redundancy forced us to eliminate a few of the checked articles. For the same reasons we have eliminated some sections of certain chapters, omissions indicated in the text by ellipses in brackets.

The general introduction and the specific introductions to each of the three sections are the joint effort of both editors. We are grateful for the assistance, support, and encouragement of Susan Herrnstein throughout the planning, the editing, and the production of this volume. We would also like to thank the Bradley Foundation for transferring funds originally granted for the preparation of Herrnstein's book to the present one. We are grateful to Will Vaughan for preparing the index. Finally, we thank the Russell Sage Foundation and the editors and staff at Harvard University Press for their dedication to the project.

The Matching Law

Introduction

Herrnstein's matching law is a theory of choice; it views life generally in terms of choice. According to the matching law, however, a choice is neither an internal decision nor an isolated output of an internal decision process; rather, a choice is a *rate* of overt events strung out over time. The matching law predicts not whether you will eat, drink, sleep, play, or work at this very moment but the fraction of hours per day you will spend eating, drinking, sleeping, playing, or working; not whether the pitcher will throw a fastball, curve, slider, or changeup at this moment but the proportion of each type of pitch he will throw to a given batter. In the introduction to the first article included here, Herrnstein states the aim of several of the essays: "to elucidate the properties of relative frequency of responding as a dependent variable." The matching law says that if an interval of time may be divided into more than one alternative activity (that is, if behavior is free to vary), animals (nonhuman and human alike) will allocate their behavior to the activities in exact proportion to the value derived from each.

In the laboratory it is possible to manipulate value experimentally by manipulating rate of reinforcement (number of food or water deliveries for hungry or thirsty pigeons or rats, or money or points exchangeable for money for human subjects). It is also possible in the laboratory to measure relative frequency of responding by observing clearly discriminable choices: a pigeon pecking one or another differently colored plastic disk; a rat pressing one or another lever; a person pressing one or another button. In all of these cases, Herrnstein found, the relative rate of responding is equal to the relative rate of reinforcement actually obtained by the subject (food or water deliveries for the pigeons, money or points exchangeable for money for the people). This equality is called matching, and the statement of the equality is called the matching law.

The first important consequence of the matching law was its laboratory extension from symmetrical choice (two disks, levers, or buttons) to asymmetrical choice, or Hobson's choice—take it or leave it (peck or do not peck a single disk; press or do not press a single lever or button). Herrnstein found that the equality between relative rates of responding and reinforcement with symmetrical choice also holds empirically when the alternatives are asymmetrical. A pigeon that would match its relative frequency of pecks on two disks to the relative frequency of reinforcement of the pecks would also match the fraction of total time available spent pecking a single disk to the fraction of total reinforcers derived from pecking the disk. (Total reinforcers derived by a pigeon in the situation include those dependent on pecking the disk, those delivered freely, independent of behavior, and those obtained through "interim" activities such as preening, wing-flapping, neck-stretching, and so on.) Previous attempts to unify symmetrical and asymmetrical choice with a single quantitative model had conceived the symmetrical case as an extension of the asymmetrical case—they started with a single response and then added another, and another. These attempts failed. For example, the best available explanation of the behavior of a rat running down a straight alley with food at the end in a "goal box" could not accurately predict the behavior of the same rat in a T-maze (in which, after running up the leg of the T, the rat chooses between the two arms, each with its own goal box at the end).

Herrnstein's extension of matching from symmetrical to asymmetrical choice provided a single theoretical umbrella for all laboratory studies of choice. Although the original proportionality has been elaborated by his students and by Herrnstein himself, the matching law remains as the most general empirical law of choice we have. If the matching law applies outside of the laboratory as well as in, and if the philosophers are right who say that ultimately all behavior of humans is choice behavior, then the matching law emerges as a general empirical law in psychology.

Many of the essays in Parts II and III extend the applicability of the matching law outside the laboratory. One area of such extension was to the behavior of nonhuman animals foraging in the wild. (An intermediate step was taken by Herrnstein's then graduate student William Baum, who placed panels with perches, plastic disks, and food hoppers in his attic. Wild pigeons could enter the attic under the eaves, perch on the perches, peck the disks, and obtain food. An editor of this volume, HR, remembers sleeping one night just below that attic and being kept awake by the clacking of the relay equipment as the wild pigeons waited their turn to confirm the empirical validity of matching.)

The matching law was extended from the laboratory to everyday human behavior, including economic behavior. The key concept behind such extension is that of "melioration," a mechanism that Herrnstein believed underlies all matching—laboratory and nonlaboratory. (Unlike the matching law itself, which is a molar law dealing with overall relative frequencies of events, melioration is a molecular mechanism that attempts to describe behavior moment by moment. If melioration is true, then the matching law must be true but the reverse is not the case. Matching could conceivably result from the action of another mechanism.)

What is melioration? Consider a person on a summer vacation who divides her time between eating and snoozing, occasionally switching back and forth between these two activities. Melioration provides a way of predicting exactly when this vacationer will make these switches. The key variables in melioration are the local reinforcement rates, which in this example are the immediate flows of value (or utils) which derive from each of the two activities (eating and snoozing). The local reinforcement rate for eating at a certain point in time is the flow of utils that the vacationer will receive during the current hour if she uses that hour to eat. Likewise the local reinforcement rate for snoozing at a certain point in time is the flow of utils that the vacationer will receive during the current hour if she uses that hour to snooze. Note that the two activities are mutually exclusive; the vacationer must choose one or the other during any given hour. Melioration posits that at any point the vacationer will choose the activity with the higher local reinforcement rate.

This choice rule implies that when all is said and done the average reinforcement rate of eating (total utils received during hours of eating divided by total number of hours spent eating) will equal the average reinforcement rate of sleeping (total utils received during hours of sleeping divided by total number of hours spent sleeping). Hence melioration implies matching. Average reinforcement rates are equalized because of diminishing returns. If the local reinforcement rate of eating is higher than the local reinforcement rate of sleeping, then melioration implies that the next unit of time will be allocated to eating, thereby driving down the local reinforcement rate of eating. In most circumstances this mechanism equalizes average reinforcement rates across alternatives.

Melioration differs in principle from the choice rules postulated by economic optimization. Economic optimization predicts that the vacationer will allocate her time to maximize total utils. Such maximization requires forward-looking decision rules, which take into account the consequences that this hour's activity choice has on all future utility flows. By contrast, a meliorating vacationer will simply choose the activity with the higher utility flow during the current hour (that is, the activity with the higher local reinforcement rate). For example, if eating this hour generates aversive heartburn the next hour—whatever activity the vacationer chooses during that next hour—it may be optimal for the vacationer to avoid eating during the current hour. However, melioration predicts that the vacationer will eat during the current hour if the local reinforcement rate of eating (which doesn't take into account the next hour's heartburn) is higher than the local reinforcement rate of snoozing. Hence melioration ignores the forward-looking considerations necessary for economic optimization.

In many instances there would be no practical difference in behavior between the economic theory and melioration. But in a series of ingenious laboratory experiments (described in several of the chapters below) Herrnstein and his students have shown that often, when the two theories predict different behavior, subjects behave as melioration predicts. The issue is important because the economic mechanism would maximize overall value in the situation while melioration, to the extent that it differs from the economic theory, also differs from maximization. There is some current dispute about the interpretation of these experiments, but their importance is unquestioned because they lead to an explanation of apparently irrational behaviors such as addiction. Irrational-seeming behavior creates severe problems for economic optimization theory which says that subjects always maximize (always choose the best of all possible alternatives). Melioration theory, precisely because it rejects optimization, accounts more easily than optimization theories do for apparently irrational behavior.

Only a few years have gone by since the initial publication of the last of these essays. Where does this body of work stand in the context of the history of thought about human and nonhuman behavior and what will its effect be on the future of psychology and economics?

Prior to the matching law there existed two main principles of analysis in psychology: analysis by reflexes and analysis by reason. Both stem from the view of human nature expressed by René Descartes in his *Treatise of Man,* originally published in 1662. According to this view all nonhuman animal behavior is governed by reflexes, stimuli from the environment mechanically causing responses. An example would be the mechanism by which a dog might pull its paw away from a fire: the heat from the fire sets up movement in nerves in the dog's paw; the nerves transmit this motion to the dog's brain; the brain in turn releases a control signal (a pneumatic signal in Descartes' model) that upsets the balance of muscular tension in the dog's leg and flexes the limb. According to Descartes, any movement of any nonhuman animal as well as many human movements could be explained in this way. The mechanism by which a boy pulls his hand away from a fire works just like the one that controls the dog's paw, Descartes believed. However, he also believed that, in addition to their reflex systems, each human possesses a soul capable of logical reasoning, a soul that interacts with, that may dominate or be dominated by, human reflexes. (For Descartes this interaction occurred in the pineal gland of the human brain.) Although some human behavior could be analyzed into just reflexes—that is, immediate environmental or bodily causes could be discovered for them—some human acts were said to have no *external* cause; their cause was said to be the *internal* rational soul, exerting its will on the body. The soul could activate the brain's pneumatic control system by itself or could suppress and counteract the force of reflexes (as when you suppress a cough or a sneeze at the theater). Self-control, a concept that Herrnstein was to put at the core of his system of behavioral analysis (see Part II), was considered by Descartes to be the soul's dominance over brute reflexes.

Despite Descartes' insistence on the power of the human soul, many subsequent investigators came to believe that all human behavior could be analyzed in terms of reflexes alone. Among these were the Russian physiologists I. M. Sechenov and I. P. Pavlov and the American behaviorists J. B. Watson, C. Hull, and B. F. Skinner (Herrnstein's teacher at Harvard). Skinner differed from his predecessors by viewing Cartesian reflexes not as fundamentally physiological but as fundamentally behavioral—as correlations between environmental stimuli and the behavior of the organism as a whole. Skinner saw reflexes as methods of controlling behavior of oneself as well as of other people. If you want a given response, all you need to do is to provide (or expose yourself to) the stimulus. Discovery of how the nervous system does its work is a task for biology, Skinner believed, not for psychology. But Skinner agreed with Descartes that something more than correlation between responses and immediate antecedent stimuli is necessary to explain the behavior of organisms. For Skinner this "something more" was the correlation between behavior and its environmental consequences. Such correlation (where the environment acts after rather than before the act) could be used to control behavior—of nonhumans as well as humans. This sort of behavioral control (by consequences) proved to be much more powerful and at the same time much more precise than the control achieved with the stimulus-response, Cartesian, type of reflex. For example, the dog might withdraw its paw in response to a painful stimulus like a fire or an electric shock. This would be a Cartesian (or S-R) reflex. But you could train the dog to shake hands, wave goodbye, or press a lever at a precisely defined rate by following these acts (or successive approximations to them) with a reward such as a dog biscuit or a pat on the head and the words, "good dog." Moreover, if only you provided these rewards or if you provided them only in a given room of the house, the dog would soon come to shake hands, wave goodbye, and so on only to you or only in that room. The complex human interactions of everyday life, Skinner felt, were nothing but elaborations of this basic process. The fundamental units of behavior that could be controlled in this way, Skinner called "operants." The principles of such control, he called "operant conditioning" to distinguish them from those of Cartesian-reflex control ("respondent conditioning"). But it is important to note that Skinner initially saw an operant as merely another type of reflex. Even later, when he abandoned this notion, the concept of a single isolated act (like a pigeon's peck on a plastic disk or a rat's lever press), an act that could be strengthened or weakened by its immediate consequences, remained the bedrock of his system.

Thus respondents, acts caused by antecedent stimuli clearly originating in the external environment (like the dog's withdrawal of its paw from the fire), were explained by Skinner in essentially the same way as they were explained by Descartes—as reflexes. But Skinner's explanation of acts without any clear antecedent cause was different from that of Descartes. Such acts were explained by Skinner not in terms of responses to an internal "will" (an aspect of the internal soul) but in terms of their consequences—in terms of what Skinner called "contingencies of reinforcement." The boy who pulls his hand from the fire is responding to a clear and distinct external antecedent stimulus. The hotter the fire, the quicker the movement. But there is no clear and distinct stimulus preceding the boy's drawing of water from the well or waving goodbye or shaking hands or writing a letter or doing arithmetic. These acts, explained by Descartes as caused by the soul, were explained by Skinner in terms of their consequences—the water, the wave back, the social approval, the completed letter, the completed sum. In the last two cases, where the immediate consequence of the behavior is not clearly a reward,

Skinner felt obligated to trace its connection (by operant or respondent chains) to an event that clearly is a reward. Most of Skinner's later works were efforts to do this. Once such connections are traced there is no need, according to Skinner, for concepts associated with the soul such as thoughts, emotions, desires. All behavior previously explained with the use of such concepts could be explained, according to Skinner, in terms of operants and their consequences—a kind of reflex. This was the method of analysis of behavior in which Herrnstein was trained. The essays in Part I begin with this Skinnerian approach to behavioral analysis but soon find it to be inadequate and reject it. What is the problem with analysis by reflex?

Analysis by reflex, like many natural laws, is basically a system for keeping accounts. Consider Helmholtz's First Law of Thermodynamics. It says that if the energy in a system at time A equals the energy in the system at time B, then between times the energy going into the system must equal the energy going out. In an automobile engine, for instance, the chemical energy going in (in the gasoline) must equal the kinetic energy going out (in the movement of the car). Otherwise, if we find a difference, we have to look for other sources of energy (heat lost, sound energy, chemical energy in the exhaust gasses, and so on). This accounting system has been so successful that, in cases where the most assiduous search fails to find a balance, physicists have invented hypothetical kinds of energy.

Analysis by reflex is just such an accounting system. If a response is observed, then (for Skinner) either a stimulus or an immediate consequence (a reinforcer) must be discovered to account for it. If no stimulus or reinforcer is apparent, one must be hypothesized. For example, a boy may avoid a painful burn by pulling his hand away from a stove, not at the touch of the flame but at the sight of the burner valve turned to "ON." The boy's hand withdrawal may be seen not as an escape from an aversive external fire but as escape from an aversive *internal* fear (learned through past conjunctions of "ON" burner valves and hot fires). In an influential article, "Method and Theory in the Study of Avoidance" (*Psychological Review*, 1969, 76, 49–69), Herrnstein argued against the use of *internal* fear as an explanatory mechanism to substitute for *external* painful stimuli. Even when no painful fire is present *at the moment*, Herrnstein argued, the causal factor in avoidance is still the general high rate of fires themselves in the presence of an ON burner. By pulling back his hand whenever the burner is ON the boy generally decreases the rate of burns from fires even though he may be accomplishing nothing at the present moment. The simplicity gained by explaining behavior in terms of its effect on the overall rate of events (such as fires) more than compensates, Herrnstein argued, for the loss of the one-to-one correspondence of stimulus and response or response and reinforcer demanded by reflex-based analyses.

In other words, Herrnstein's analysis of avoidance responding proposed that we look for the causes of behavior not *deep* within the organism but *widely,* in time, in the contingencies between the frequency of an act and the overall rate of significant events (such as fires, food, sexual opportunities) in the external world.

Herrnstein's matching law was another point in this same argument against reflex-based theories. First, the matching law's units are *rates* of behavior and *rates* of reinforcement rather than individual acts and individual consequences. Second, the matching law puts a strain on reflex-based analysis by showing that a given act may be manipulated through the consequences of *other* acts. A person might buy more potato chips if they were made to taste better (increasing the reinforcing power of potato chips themselves). So much is consistent with a reflex-based analysis of behavior. But the person might buy just as many more potato chips if pretzels were made to taste worse (decreasing the reinforcing power of alternatives). A reflex-based analysis cannot easily account for indirect behavioral manipulation. The train of internal and external events that would be needed to bridge between the increase or decrease of reinforcement of response *B* and the inverse variation of response *A* was just too cumbersome (and unsupported by evidence) to maintain. The chapters in Part I present evidence of indirect manipulation of behavior. They also present Herrnstein's alternative analysis in terms of the matching law which focuses not on the consequences of an act itself but on the consequences of that act relative to the consequences of other acts. Herrnstein's relativistic analysis turned out to be much more simple, more straightforward, and more in accord with the data than Skinner's individual-response analysis. And the relativity itself broke clearly with Cartesian reflexes. If you could manipulate one act by reinforcing another then neither could be a reflex in any sense.

The other main form of behavioral analysis initiated by Descartes is analysis by reason. What sort of behavioral analysis may be based on reason? Whereas, after Descartes, analysis by reflex alone was pursued and expanded by physiologists, analysis by reason was pursued by philosophers. In modern versions of analysis by reason, rational behavior is not necessarily consciously rational. A person may have a smoothly operating and perfectly logical internal information-processing apparatus without necessarily being aware of its operation. Perhaps the purest modern scientific example of analysis by reason is in microeconomics which holds that individual consumers maximize utility: that is, of all possible (economic) alternatives among which a person can choose, the person will consistently make the best choice, the one that (given the information available and other constraints on behavior) "maximizes utility." How this maximization or optimization is accomplished is a matter of debate but *that* it is accomplished is generally accepted among economists. But to say that behavior maximizes utility is to say that behavior is rational (although not necessarily consciously so). To the extent that behavior does not maximize utility—to the extent that a person could have made a better choice—that person's behavior must not be rational.

Some economists (for example, Gary Becker and Kevin Murphy, 1988) have extended economic theory into the realm of psychology, to cover behavior such as addiction and habits, behavior not ordinarily conceived as rational. Despite appearances, these economists say, even addicts are making rational choices—the economists' theories are attempts to derive maximization-based models that predict those choices. These economists first attempt to express a person's underlying preferences in terms of utility functions which are then used to predict future behavior. To take a notoriously common example, the alcoholic who says, "I wish I weren't an addict," but who nonetheless remains addicted, may be seen as rational if we consider the possibility that the transition path from alcoholism to sobriety is extremely aversive. The alcoholic would like to be at the end of that path (that is, sober), but getting there is so costly (in terms of lost utility), that the alcoholic continues to drink.

The chapters in Part III consider and (partially) reject this economic view, this analysis by reason. What was it that Herrnstein found unsatisfactory about analysis by reason?

Like analysis by reflex, optimality analysis (analysis by reason) is basically a method of keeping accounts. The most valuable alternative must always be chosen, the economist says. If a person (or a nonhuman animal) appears to choose a less valuable over a more valuable alternative, the optimality theorist must either recalculate value or reconsider how choice may have been constrained; it is the theorist, the economist says, not the chooser, who has made a mistake. Just as the physicist must recalculate energies when the First Law of Thermodynamics appears to be violated, so the economist must recalculate values or reestimate constraints when people appear to violate the principle of optimality. In economics value is given by a utility function (utility as a function of all possible choice alternatives) which both summarizes past behavior and predicts future behavior. The utility function is a means of assigning a value to each alternative; then choice must be such as to obtain the highest valued alternative. In other words, optimality theory says that organisms always behave (within the constraints of available alternatives) to maximize utility. Unlike analysis by reflex, this sort of analysis by reason is clearly relativistic and in that respect consistent with Herrnstein's empirical findings: a person who always chooses the highest-valued alternative would choose X if Y were less valuable than X but reject X if Y were more valuable than X. It appeared at first (and still appears to many theorists) that the matching law is consistent with optimality. Much of the research reported in Part III, however, is an attempt to separate these two principles. Herrnstein asks: is the obtained behavioral allocation best described as the outcome of melioration or as the outcome of economic optimization? Although by judicious postulation of value and constraint, either description (either accounting method) may be applied to any observed allocation, Herrnstein's experimental results are readily analyzable in terms of matching and melioration but, only with some difficulty, are they analyzable in terms of optimization. What Herrnstein found unsatisfactory about analysis by reason (as epitomized by economic optimization theory) was essentially the difficulty in stretching it to fit the data presented in these chapters.

In Herrnstein's laboratory experiments (unlike natural situations) the structure of the alternatives and their values are made clear and distinct (just as in thermodynamics experiments the energy inputs and outputs are made clear and distinct). In the laboratory, where crucial parameters may be manipulated, matching (and equality of local rates) is a general empirical finding. The real world of human and nonhuman choice (like the real world of woodstoves and automobiles) contains fuzzier boundaries. There, the matching law (like the First Law of Thermodynamics) is an assumption, a highly useful method of keeping accounts, which allows us to predict, control, and understand nature.

Part I

The Matching Law:
Against Reflexology

The four essays in this section progress from Herrnstein's first tentative statement of the empirical matching law, to a marshalling of evidence in support of the extension of matching from symmetrical to asymmetrical choice, to a pair of derivations of matching from more fundamental principles.

Chapter 1 reports Herrnstein's initial discovery of matching in the context of a study of the behavior of pigeons exposed to concurrent variable-interval (VI) schedules of reinforcement. The variable-interval schedules follow a standard procedure developed by Skinner for operant research. The VI control mechanism arranges for a reinforcer (a small amount of grain, in this case) to be delivered to a deprived animal (a hungry pigeon) immediately after a response (a peck on a disk). But not every response is reinforced. The schedule of reinforcement determines which responses will be reinforced. Under a variable-interval schedule, an interval of time (variable from interval to interval) must elapse since the last reinforcer was delivered. Responses during this interval are not reinforced. After the interval is over (a point unsignaled to the pigeon) the next response is reinforced. Two variable-interval schedules may differ in the average of these (inter-reinforcement) intervals. A VI 30-second schedule, for instance, programs reinforcers on the average every 30 seconds; a VI 60-second schedule programs reinforcers on the average every minute. (However, there will be some long inter-reinforcement intervals in the VI 30-second schedule and some short intervals in the VI 60-second schedule.)

The VI schedule is analogous to the everyday-life testing of a piece of food for doneness as it sits in the oven (when you have no idea at all of how long it will take). The food could be ready in a minute, an hour, or several hours. The only way to know is to test it by, say, cutting off and tasting a piece. The pecking response of the pigeon is essentially such a testing response. (By pecking the disk the pigeon asks the question, Has the scheduled interval elapsed?)

A variable-interval schedule of reinforcement generates a rather steady slow rate of responding (of pecking). In Chapter 1 Herrnstein shows

what happens when two different variable-interval reinforcement schedules are programmed concurrently (the pigeon pecks two disks located
on the wall of the Skinner-box; the plastic disks look like two illuminated elevator buttons lined up side by side instead of vertically). The
experiment uses a changeover delay (COD), a procedure that penalizes
frequent switching (alternation from disk to disk) by delaying reinforcement for a brief period after each switch. The COD essentially commits
the subject to abide briefly to a choice once it is made. (In analogous experiments with humans pressing buttons the COD is often augmented or
replaced by a device that prevents subjects from pressing both buttons at
once.)

Figure 1.1 shows the matching relationship. (The pigeons distributed
their pecks on the two disks proportionally to the reinforcers obtained
from each.) At this point Herrnstein was very tentative about the generality of matching. The matching equation appears here (in 1961) only as
a particular relationship between pecking and eating by pigeons: the rate
of pecking on one key (p_1) divided by total pecking ($p_1 + p_2$) equals the
rate of eating contingent on pecking one key (e_1) divided by total eating
($e_1 + e_2$).

In Chapter 2, published fifteen years later, Peter de Villiers and Herrnstein, on the basis of "approximately 40 experiments on rats, monkeys,
pigeons, and in one case, human beings," confidently expand the elements of matching from pecks to responses generally, from eating to
reinforcement generally, and from two alternatives to any number of
alternatives. Why then was such a simple relationship not already well
established in 1961? In retrospect there seem to be two answers to this
question. First, and the theme of Chapter 2, is that while the frequency
of single responses (studied in so-called nonchoice procedures) seems
not to be relevant to matching law analysis, a few assumptions (simple in retrospect) subsume such studies under the matching law. De Villiers and Herrnstein are saying that all behavior is choice behavior. The
(asymmetrical) choice between wearing and not wearing a coat, they
say, is just as well described by the matching law as the (symmetrical)
choice between wearing a red coat and wearing a green coat. The crucial assumption in the asymmetrical case is that *not* doing activity X
may also be reinforced and that this other reinforcement may have a
strong effect on the frequency of activity X. In choosing to do activity X we are forgoing all other activities that might have taken up the
same time. In other words, the total amount of behavior (like the total amount of time) in a situation is constant; reinforcement just deter-

mines how we carve it up. (Chapter 3 derives the matching law from this principle acting on a molecular level.) Previous behavioral studies of single responses had varied only the reinforcement of specific acts. The matching law (Equation 2.1) agrees with previous (reflexive) accounts that reinforcing a response increases its frequency. But the matching law goes on to describe the functional form of that increase (see, for instance, Figure 2.1) and, more interestingly, asserts that any other reinforcement in the situation—whether it consists of extrinsic reinforcement of another response, freely delivered extrinsic reinforcement independent of any response, or intrinsic reinforcement of activities (such as pigeons' preening or humans' playing) not obtainable by making the response itself—*decreases* the frequency of the response. This fact, like most important facts, is fairly obvious when you think about it. Yet its incorporation into a principle of choice awaited Herrnstein's matching law.

The second reason that the matching law remained undiscovered is that many choices of everyday life and virtually all choices previously studied in the laboratory (the rat in the maze, for instance) involve an interaction that, while obeying the matching law, obscures the fact of matching. For example, if someone repeatedly offers you a choice between $1 and $2, you will (presumably) take the $2 every time. In a trivial sense, this is matching. Your relative frequency of taking the $2 alternative (100%) equals the relative frequency of $2 outcomes received (100%). Or, if both alternatives yielded only $1 but selecting one required you to reach out and take it while selecting the other required lifting up a heavy weight, you would presumably choose the less effortful alternative 100% of the time. Or, if you were choosing between a gamble with 3:1 odds against winning and one with the same payoff but 4:1 odds, you would presumably always choose the higher probability payoff (the lower odds). All of these situations yield exclusive choice of one alternative and matching only in a trivial sense. (The only reason you might distribute choices over both outcomes in such situations would be if the outcomes were not perfectly discriminable or if, as often in life, they were changeable and the only way to discover whether an outcome had changed was to choose it occasionally.)

Chapter 4 presents Herrnstein's melioration mechanism and shows how it would produce exclusive choice of the higher alternative in such situations. Exclusive choice of the higher-valued alternative would maximize total reinforcement as well as match. In this essay Herrnstein goes on to discuss other situations in which maximizing and melioration deviate in their predictions; in all cases melioration predicts observed choice while subjects fail to extract the highest possible overall reinforcement rate from the situation (they fail to maximize overall reinforcement rate).

1 Relative and Absolute Strength of Response as a Function of Frequency of Reinforcement

In Herrnstein (1958) I reported how pigeons behave on a concurrent schedule under which they peck at either of two response-keys. The significant finding of this investigation was that the relative frequency of responding to each of the keys may be controlled within narrow limits by adjustments in an independent variable. In brief, the requirement for reinforcement in this procedure is the emission of a minimum number of pecks to each of the keys. The pigeon receives food when it completes the requirement on both keys. The frequency of responding to each key was a close approximation to the minimum requirement.

The present experiment explores the relative frequency of responding further. In the earlier study it was shown that the output of behavior to each of two keys may be controlled by specific requirements of outputs. Now we are investigating output as a function of frequency of reinforcement. The earlier experiment may be considered a study of differential reinforcement; the present one, a study of strength of response. Both experiments are attempts to elucidate the properties of relative frequency of responding as a dependent variable.

Originally published in *Journal of the Experimental Analysis of Behavior*, 1961, 4, 267–272. Copyright 1961 by the Society for the Experimental Analysis of Behavior, Inc. The work reported in this essay was supported by Grant G-6435 from the National Science Foundation, Washington, D.C., to Harvard University. The author wishes to express his indebtedness to Dr. Douglas Anger of the Upjohn Company for his valuable comments concerning the interpretation of the data in this experiment.

Method

Subjects

Three adult, male, White Carneaux pigeons, maintained at 80% of free-feeding weights, and experimentally naive at the start of the study, were used.

Apparatus

A conventional experimental chamber for pigeons (Ferster and Skinner, 1957) was modified to contain two response-keys. Each key was a hinged, translucent Plexiglas plate mounted behind a hole in the center partition of the chamber. The pigeons pecked at a circular area (diameter = 0.75 inch) of the plate, and a force of at least 15 grams was necessary to activate the controlling circuitry. Any effective response operated an audible relay behind the center partition; it has been found that the resulting auditory feedback stabilizes the topography of pecking. Behind each key was a group of Christmas-tree lamps of various colors, each group mounted in such a way that it cast significant amounts of light through only one key. The two keys were 4.5 inches apart (center-to-center) around the vertical midline of the center partition and on a horizontal line about 9 inches from the floor of the chamber. Through a 2-inch-square hole in the center partition, 2 inches from the floor, the pigeon occasionally received the reinforcer—4 seconds' access to grain.

A masking noise and a low level of general illumination were provided.

Procedure

Preliminary training lasted for two sessions of 60 reinforcements each. During these sessions, a peck to either key was reinforced only when the just-previous reinforcement was for a peck to the other key. This alternating pattern of reinforcement led rapidly to a pattern of responding that consisted of almost perfect alternation between the two keys. The left key was always red; the right, always white.

During the experiment proper, responding to either key was reinforced on a variable-interval schedule. The schedule for one key was independent of the schedule for the other. Thus, at any given moment, reinforcement could be made available on neither key, on one key or the other, or on both keys. A reinforced response to one key had no effect on the programmer that scheduled reinforcements on the other.

The primary independent variable was the mean time interval between reinforcements on each key. These intervals were chosen so that the mean interval of reinforcement for the two keys taken together was held constant at 1.5 minutes.[1] The overall average value of 1.5 minutes was produced by a number of pairs of values for the two keys. The combined frequency of reinforcement from independent variable-interval schedules will be a constant if the values for each of the two keys are chosen according to the hyperbolic relationship:

$$\frac{1}{x} + \frac{1}{y} = \frac{1}{c};$$
                                                                                                    (1.1)

in which x is the mean interval on one key, y is the mean interval on the other, and c is the combined mean interval for the two keys taken together. The pairs of values used were VI(3) VI(3); VI(2.25) VI(4.5); VI(1.8) VI(9); and VI(1.5) VI(∞)—i.e., extinction on one of the keys.

During most of the experiment, the pigeons were penalized for switching from one key to the other. Each time a peck to one key followed a peck to the other key, no reinforcement was possible for 1.5 seconds. Thus, the pigeon never got fed immediately after changing keys. When the pigeon switched keys before the 1.5-second period was completed, the period simply started anew. At least two consecutive pecks on a given key were necessary before reinforcement was possible: the first peck to start the period, and the second after it was completed. This penalty for alternation will be referred to as the "change-over delay of 1.5 seconds," or COD (1.5″).

Sessions lasted for 60 reinforcements, which required approximately 90 minutes since the overall mean interval of reinforcement was always 1.5 minutes.

Results

Figure 1.1 shows the relative frequency with which the pigeon pecked on Key A [the left key] as a function of the relative frequency with which it

---

1. It should be noted that, by convention, the mean of a variable-interval schedule refers to the *minimum* average inter-reinforcement time and not to the *actual* inter-reinforcement time obtained under conditions of responding. Thus, if a particular animal responds very slowly, the actual mean interval of reinforcement may be larger than the value designated by the experimenter. The value designated is a minimum that is closely approached in practice because the animal's rate of responding is ordinarily high in comparison to the intervals in the reinforcement schedule.

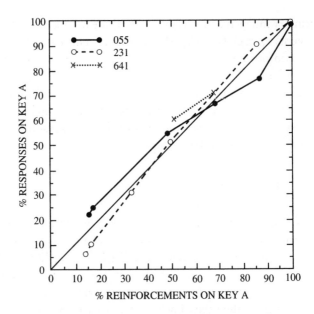

Figure 1.1 Relative frequency of responding to Key A as a function of relative frequency of reinforcement on Key A, for three pigeons. COD (1.5″) is present throughout.

was reinforced on that key. Each point on the graph is a mean of the last five sessions under a given pair of values of the variable-interval schedule. The COD operated on all these sessions. The ordinate and abscissa values were calculated by comparable methods. The number of responses (ordinate) or reinforcements (abscissa) on Key A was divided by the total number of responses or reinforcements, respectively. The five last sessions were pooled to make this computation.

The diagonal line with a slope equal to 1.0 in Figure 1.1 shows the function that would be obtained if the relative frequency of responding were exactly equal to the relative frequency of reinforcement. The empirical values approximate the theoretical function with a maximum discrepancy of only about 8%. There seems to be no regular pattern to the deviations from the theoretical function.

The absolute rate of responding on each of the keys is shown in Figure 1.2. Responses per hour are plotted against reinforcements per hour, for each key separately and for the two pigeons (231 and 055) that had an appreciable range of the independent variable. Data from the same sessions are plotted in Figures 1.1 and 1.2. With one exception (Pigeon

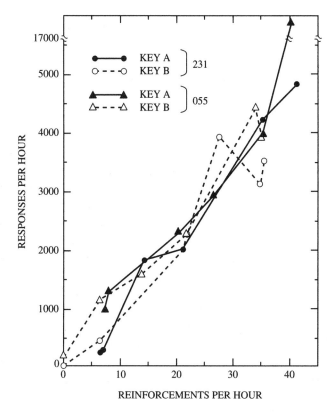

Figure 1.2   Rate of responding on each key as a function of rate of reinforcement on that key for two pigeons. COD (1.5″) is present throughout.

055, Key A, at 40 reinforcements per hour), the points in Figure 1.2 approximate a linear function that passes through the origin. It will be shown later that this relation between absolute rate of responding and absolute rate of reinforcement is the simplest one that is compatible with the relative-frequency function presented in Figure 1.1. [. . .]

Discussion

The major problem posed by the present experiment is to explain the simple correspondence in Figure 1.1 between the relative frequency of reinforcement and the relative frequency of responding. In a sense, this correspondence is readily explained by the curves in Figure 1.2, which suggest that the relation between the absolute rate of responding and the absolute rate of reinforcement is a linear function that passes through the origin. If this relation is represented as $p = ke$, in which p and e denote the absolute frequencies of pecking and eating, then the simple matching function of Figure 1.1 may be expected to follow the form

$$\frac{p_1}{p_1 + p_2} = \frac{ke_1}{k(e_1 + e_2)}. \tag{1.2}$$

The constant, k, drops out and the remaining expressions on each side of the equation denote relative frequencies of responding and reinforcement. The equality of these two relative frequencies may thus be regarded as a consequence of a linear relation, of any slope and zero intercept, between the absolute frequencies. Moreover, this relation between the absolute rates of responding and reinforcement is one that is consonant with a plausible view of response strength: Rate of responding is a linear measure of response strength, which is itself a linear function of frequency of reinforcement. The correspondence in Figure 1.1 would thereby result from the fact that the behavior on each of the two keys obeys a simple linear rule governing strength of response. According to this point of view, the animals match relative frequency of responding to relative frequency of reinforcement not because they take into account what is happening on the two keys, but because they respond to the two keys independently.

The critical relation, $p = ke$, has been asserted before. Skinner (1938, p. 130) has discussed a quantity called the extinction ratio, which is the total number of responses divided by the number of reinforced responses in a fixed-interval schedule of reinforcement.[2] He presented a small amount of data that indicated that this quantity remained constant as the size of the fixed interval was varied. The constancy of the extinction ratio is merely another form, $p/e = k$, of the function we find.

Perhaps the greatest vulnerability of the foregoing account lies in its simplicity. If it were true that the rate of responding is so simply related to the frequency of reinforcement, the fact ought to have been well established by now. We should expect that behavior in a single-key situation would reveal the same linear relation shown in Figure 1.2, and that with

---

2. Skinner defined the extinction ratio as the number of unreinforced responses divided by the number of reinforced responses, but in actual computation he used the total number of responses divided by the number of reinforced responses. The difference is of no significance for the present discussion since both definitions imply a linear relation with zero intercept between absolute rate of responding and absolute rate of reinforcement.

all the work done with the single-key problem, the nature of the relation between rate of responding and frequency of reinforcement would be known. Unfortunately, this information is not available. In few studies has the frequency of reinforcement been varied over an adequately wide range. Those which have done so have usually also involved manipulations in other, and possibly contaminating, variables. [. . .]

The suggestion of the present discussion is that the surprisingly precise correspondence between relative frequency of responding and relative frequency of reinforcement arises from the function relating absolute frequency of responding and absolute frequency of reinforcement. When this function is linear with an intercept of zero, matching is found. In single-key situations, this linear relation is not obtained; and it is also not obtained under concurrent schedules unless some additional procedural factor reduces the pigeon's tendency to over-respond at low frequencies of reinforcement and under-respond at high. The COD is such a procedural factor; but others, such as distance between keys or effort involved in the response, may also be satisfactory. The duration of the COD may or may not be critical in the effect it has on the slope of the relative-frequency function. If a broad range of durations of the COD all give approximately perfect matching, then it seems correct to say that the concurrent procedure is a good one for studying absolute, as well as relative, strength of responding. In single-key situations, the rate of responding is not very sensitive to frequency of reinforcement. This insensitivity probably weakens our interest in the concept of strength of response. It may be that the concept can be given significant empirical support in multiple-key situations.

Summary

A two-key, concurrent procedure involving variable-interval schedule on each key was used. The value of the mean interval on each key was varied over a range from 1.5 to 9 minutes, but the total frequency of reinforcement for the two keys taken together was held constant. The pigeon was penalized for alternating in response between the two keys by making reinforcement impossible for 1.5 seconds after every alternation. It was found that the relative frequency of responding on a given key closely approximated the relative frequency of reinforcement on that key.

2

Toward a Law of Response Strength

PETER A. DE VILLIERS AND RICHARD J. HERRNSTEIN

Confronted with choices differing only in the frequency of reinforcement, subjects match the distribution of response alternatives to the distribution of reinforcements (de Villiers, 1977; Herrnstein, 1970, 1971). That is to say, the choices come to obey the following equation, in which B_1 to B_n enumerates the response alternatives and R_1 to R_n the reinforcements associated with each:

$$\frac{B_1}{B_1 + B_2 + B_3 + \cdots + B_n} = \frac{R_1}{R_1 + R_2 + R_3 + \cdots + R_n}. \qquad (2.1)$$

Equation 2.1, sometimes called the matching law, says that the relative frequency of each kind of responding equals, or matches, the relative frequency of its reinforcement.[1]

Originally published in *Psychological Bulletin*, 1976, *83*:6, 1131–1153. Copyright © 1976 by the American Psychological Association. Reprinted with permission. The preparation of this essay and the new data reported in it were partially supported by National Science Foundation Grant BMS74-24234 to Harvard University and National Institute of Mental Health Grant MH-15494 to Harvard University. Parts of this essay overlap with parts of de Villiers (1977).

1. For those not acquainted with this literature, a couple of elementary distinctions are in order. The matching law, first of all, deals with *obtained* reinforcement frequencies, not programmed frequencies. Thus, the "independent" variable is to some degree dependent on the subject's behavior. In the laboratory settings in which the law has been tested, this distinction is largely academic. However, in practical settings, in which response frequencies can often materially influence reinforcement frequencies, the distinction becomes crucial. The law appears to apply quite generally (e.g., Baum, 1974; de Villiers, 1977), *if* obtained reinforcement frequencies are used. Second, the matching law expresses a function between simple frequencies, or rates, of responding and reinforcement, not probabilities. Thus, the matching law differs from what has been called (Bitterman, 1965) "probability matching," although it can be shown that the two are formally related (Herrnstein, 1970; Herrnstein and Loveland, 1975). Confronted with choices that differ in reinforcement *probability*, subjects may maximize by choosing the larger probability exclusively. But maximization is precisely consistent with Equation 2.1, since Equation 2.1 refers to obtained reinforcement frequencies (Herrnstein and Loveland, 1975).

A large experimental literature supports the matching law. Although some workers have proposed other formal representations (Baum, 1974; Catania, 1973; Killeen, 1972; Rachlin, 1971; Staddon, 1972), there is no doubt that symmetrical choices in numerous species including human subjects are well described by Equation 2.1. As long as the alternative responses are equivalent in topography and the reinforcements vary only in frequency (or some other clearly quantitative parameter, such as amount of reinforcement), the distribution of choices is well predicted by the associated distribution of reinforcements (see de Villiers, 1977, for a discussion of exceptions). When the responses vary significantly in form or the reinforcements vary in quality or type (such as apples versus oranges or sex versus serenity), the equation calls for reinterpretation, at the least (e.g., Brown and Herrnstein, 1975; Rachlin, 1971).

This chapter, however, is not primarily about Equation 2.1 but about the motivational principle that is implicit in it, by which we mean simply a principle that concerns sheer *amounts* of behavior. Thus, while Equation 2.1 predicts choice, not quantity of behavior, our position is that choice is merely behavior in the context of other behavior, *not* a distinctive psychological process of its own. Typically, choice is measured as a behavior ratio (Tolman, 1938)—a frequency of one type of behavior divided by the frequency of some other type or by the total frequency of all types under observation, as is the case in Equation 2.1. If choice obeys Equation 2.1, it therefore follows from our outlook that absolute response frequencies should obey a suitable derivation of Equation 2.1. This chapter mainly reviews the evidence linking the principle of choice in Equation 2.1 to data on absolute frequencies of behavior. In a sense, this reverses the more typical progression in science from simple to complex. However, it should be evident later in the chapter why the apparently more complex phenomena of choice were formalized sooner than the more elementary (if not simple) phenomena of sheer response output.

Earlier work on the matching law (Herrnstein, 1970) has shown that the simple rate of responding to each alternative in a continuously available choice is governed by the following equation:

$$B_1 = \frac{k_1 R_1}{\sum\limits_{i=0}^{n} R_i}. \tag{2.2}$$

The rate of responding (i.e., B_1, a frequency per some period of observation) is proportional to the *relative* frequency of reinforcement i.e., $\left(\dfrac{R_1}{\sum\limits_{i=0}^{n} R_i} \right)$.

Note that the denominator here is the total of *all* reinforcement being obtained in the observation period. The constant of proportionality, k_1, is a parameter specific to each form of responding, hence its subscript. A discussion of the formal properties of Equation 2.2 is given in Herrnstein (1974). Equation 2.2 follows from Equation 2.1. To see how, let us consider an organism with just two forms of behavior in its repertoire, B_1 and B_2, each obeying the simple matching law (Equation 2.1). By cross multiplication, the frequency of each response is:

$$B_1 = (B_1 + B_2)\frac{R_1}{R_1 + R_2} \tag{2.3a}$$

$$B_2 = (B_1 + B_2)\frac{R_2}{R_1 + R_2}. \tag{2.3b}$$

By definition, $(B_1 + B_2)$ equals the total amount of behavior, which we may label k and which reduces Equations 2.3a and 2.3b to Equation 2.2. It now remains only to be shown that k is invariant with changes in R_1 or R_2. If reinforcement for B_1 is changed by any factor a, each response will change according to Equation 2.1, as follows:

$$\frac{B_1'}{B_1' + B_2'} = \frac{aR_1}{aR_1 + R_2} \tag{2.4a}$$

$$\frac{B_2'}{B_1' + B_2'} = \frac{R_2}{aR_1 + R_2}. \tag{2.4b}$$

B_1' and B_2' are the new frequencies of responding, and R_1 and R_2 are the original amounts of reinforcement. Equations 2.4a and 2.4b thus show how relative response frequencies change when the reinforcement for B_1 is changed from R_1 to aR_1. The net change in relative response is obtained by adding the differences between the old and new proportions for each response:

$$\left(\frac{B_1'}{B_1' + B_2'} - \frac{B_1}{B_1 + B_2} \right) + \left(\frac{B_2'}{B_1' + B_2'} - \frac{B_2}{B_1 + B_2} \right). \tag{2.5}$$

But the expression in Equation 2.5 must equal zero, since

$$\frac{B_1'}{B_1' + B_2'} + \frac{B_2'}{B_1' + B_2'} = \frac{B_1}{B_1 + B_2} + \frac{B_2}{B_1 + B_2} = 1.0. \tag{2.6}$$

If the net change in the proportions of B_1 and B_2 is zero, and if (as assumed) $B_1 + B_2$ equals total behavior, then insofar as the effects of changes in reinforcement are captured by Equation 2.1, the total behavior, k, must remain invariant. Any increase or decrease in the frequency of one response is exactly matched by an equal decrease or increase in the frequency of the other response, respectively, given only two responses, B_1 and B_2.

It is trivial to show that this argument can be extended to any number of response alternatives or any change in reinforcement. In short, our theory says that relative frequencies of response obey Equation 2.1, and that this implies the rule governing absolute frequencies given in Equation 2.2 (as qualified in Herrnstein, 1974). Both equations assume only that response frequencies are being controlled by the associated quantities of reinforcement.

Let us suppose that an experiment pits two alternatives of equivalent form, B_1 and B_2, against each other. Each alternative is associated with a certain amount of reinforcement, R_1 and R_2. There is also some background level of reinforcement, R_e, which contains all the other reinforcers that a subject brings with itself or finds in the experimental setting. For the familiar pigeon or rat working in an operant-conditioning chamber, R_e contains the distractions from R_1 and R_2, the reinforcements programmed by the experimenter. Presumably, R_e would then be small relative to $R_1 + R_2$. For a human subject working for the probably feeble reinforcements of the usual psychological experiment, R_e might be substantial relative to $R_1 + R_2$. In any event, the following equations show how Equation 2.2 applied to the sheer frequency of each alternative produces the matching law with respect to the choice between the two alternatives:

1. $B_1 = \dfrac{k_1 R_1}{R_1 + R_2 + R_e}$

2. $B_2 = \dfrac{k_2 R_2}{R_1 + R_2 + R_e}$

3. $k_1 = k_2$

$$4. \quad \frac{B_1}{B_1 + B_2} = \frac{\frac{k_1 R_1}{R_1 + R_2 + R_e}}{\frac{k_1 R_1}{R_1 + R_2 + R_e} + \frac{k_2 R_2}{R_1 + R_2 + R_e}} = \frac{R_1}{R_1 + R_2}. \tag{2.7}$$

Steps 1 and 2 simply explicate Equation 2.2 for the case in which the subject has two programmed sources of reinforcement, R_1 and R_2, plus the unprogrammed, extraneous source R_e. Step 3 may need a word of explanation. We said that the two alternatives were equivalent response forms, which means, by definition, that their ks are equal (see Herrnstein, 1974). Equation 2.7 shows that Equation 2.2, a principle of sheer response output, applied to each alternative implies matching (Equation 2.1), a principle of choice. It now remains to be shown that existing data do, in fact, confirm Equation 2.2, it being well established that data on choice confirm Equation 2.1 under the stated limits.

An Equation for Simple Action

In a single-response procedure, Equation 2.2 becomes

$$B_1 = \frac{k R_1}{R_1 + R_e}, \tag{2.8}$$

where B_1 is the rate of some response and R_1 is the frequency (or quantity) of reinforcement for that response. This is essentially the standard operant procedure. A subject has a single response operandum (a lever, a button, etc.) and is reinforced for working at it. As the amount or rate of reinforcement varies, the amount or rate of behavior covaries by the function in Equation 2.8. The parameter k represents the asymptotic response rate in the absence of any reinforcement for competing responses, that is, when R_1 equals ΣR_i in Equation 2.8. The k has the same units of measurement as R_1 (i.e., behavior per unit time, such as responses per minute or running speed). The total reinforcement besides R_1 in the experimental situation is represented by R_e, which is measured in terms of the same units as R_1 (i.e., an amount of reinforcement, such as pellets per hour or percent of sucrose).

Variable-Interval Schedules

Herrnstein (1970) tested Equation 2.8 with data from pigeons on the effects of frequency of food reinforcement. The most complete data come from an exhaustive study of variable-interval (VI) schedule by Catania and Reynolds (1968). A VI schedule programs a series of irregular times between successive reinforcements. Actually, since the subject must execute a certain response (e.g., key pecking by pigeons) to collect its reinforcement, the schedule arranges a series of minimum interreinforcement times. In practice, the pigeons in this experiment (and other subjects in other experiments) worked so fast that the reinforcements were collected virtually as quickly as the schedule programmed them.

In Catania and Reynolds' experiment, six pigeons were exposed to VI schedules with frequencies of reinforcement ranging from 8 to 300 reinforcements per hour. Session durations and the size of each reinforcement were set so low that the amount of satiation was probably negligible. Unless otherwise noted, the drive levels can be assumed to be constant in the experiments summarized. Each schedule was maintained for enough daily sessions to stabilize the responding. It can therefore also be assumed that we are examining steady-state responding, not acquisition.

Figure 2.1 shows the least squares fit of Equation 2.8 to the data from each of the pigeons.[2] The percentage of data variance accounted for by the equation in each case is given in the bottom right-hand corner of each panel. With a k of 86.3 responses per minute and an R_e of 7.3 reinforcements per hour, the equation also accounts for 91.3% of the data variance when response rate is averaged across the pigeons for each VI value. The units used for B_1 and R_1 determine the values obtained in the least squares solution for k and R_e. However, changing the units (e.g., to responses per *hour* or reinforcements per *minute*) would not alter the conclusion that Equation 2.8 accounts for about 91% of the variance in the average data from this experiment. What does this mean? First and foremost, it says that Equation 2.8 captures much of the functional form of the relation between work and wages for these pigeons. Second, it says that the extraneous reinforcement during experimental sessions was equivalent in value to about 7 per hour of the food reinforcements. Finally, it says that if the extraneous reinforcement had

---

2. A digital computer iteratively calculated the mean squared deviation of obtained response rates from those predicted by Equation 2.8 for a wide range of k and R_e values varying on either side of those derived from a best fit by eye. The smallest mean squared deviation thus obtained was subtracted from the total variance in obtained response rate and the result divided by the total variance to determine the percentage of data variance accounted for by the equation.

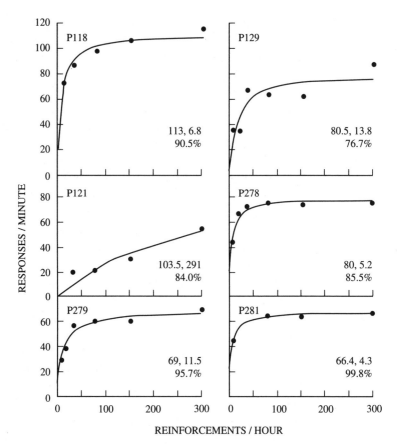

Figure 2.1   Rate of responding as a function of frequency of food reinforcement for Catania and Reynolds' (1968) six pigeons responding on single VI schedules. (The least squares fit of Equation 2.8 to the data is plotted for each pigeon. The k and R_e values and the percentage of the variance in response rate accounted for by these functions are also shown.)

been zero, the average rate of key pecking would have been about 86 per minute.

Figure 2.2 shows comparable data from an unpublished experiment by de Villiers. Three rats worked on VI schedules for food reinforcement (sweetened condensed milk). Five or six different rates of reinforcement were used, allowing the rat to reach stable rates of responding at each. Although the rats differed substantially in absolute rates of responding, each one's behavior is well described by Equation 2.8, with 83% to 99.8% of the data variance accounted for.

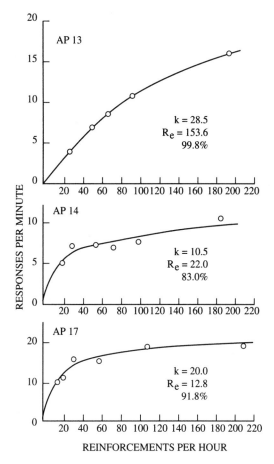

Figure 2.2   Rate of responding by three rats as a function of frequency of food reinforcement. (The smooth curves show the least squares fit of Equation 2.8 to the data. The k and R_e values and the percentage of variance in response rate accounted for by these functions are also shown. Unpublished data from P. A. de Villiers.)

Other Schedules of Reinforcement

Chung (1966) reinforced pigeons' responses on a tandem fixed-ratio, fixed-interval (tan FR FI) schedule. On this schedule, the first key peck after a reinforcement started a timer and the first response after the timer finished its cycle produced food reinforcement. Each setting of the timer yielded a certain rate of reinforcement and a resulting rate of responding. By this method, reinforcement rates of up to 2,000 per hour were briefly sustained, many times the maximum values investigated by Catania and

Reynolds (1968). Herrnstein (1970) demonstrated that Equation 2.8 also accounts for Chung's results, though with somewhat higher parameter values than those for Catania and Reynolds' subjects. With a k of 130 pecks per minute and R_e of 210 reinforcements per hour, Equation 2.8 accounts for 94.7% of the variance in mean response rate for Chung's pigeons.

The higher value for R_e may reflect various differences between Chung's experiment and Catania and Reynolds'. If Chung's pigeons were less hungry or were receiving smaller quantities of food as reinforcement, then, ceteris paribus, R_e would be larger. This follows because the extraneous reinforcement, R_e, is measured in the units of the programmed reinforcement. The smaller these units are, the larger the number of them it takes to measure a given amount of extraneous reinforcement. In fact, Chung's pigeons were food deprived to 90% of their normal weight, whereas Catania and Reynolds' were food deprived to 80%. Moreover, the reinforcement for Chung's pigeons was 3-sec access to pigeon feed, whereas for Catania and Reynolds', it was 4 sec. Both differences are in the right direction, although there remains the possibility that extraneous reinforcement was actually different in the two experiments, since they were conducted in different apparatuses. The different apparatuses may also account for the different values of k, for which we otherwise have no ready explanation. Different topographies are suggested by the variation in this parameter.

Equation 2.8 applies most directly to VI schedules, in which the probability of a reinforcement being scheduled after one has been presented is relatively constant over time. Its extension to other schedules, such as fixed-interval (FI), is more difficult. On an FI schedule, a response is reinforced only after a constant time passes since the last reinforcement. Subjects rapidly learn a time discrimination (Ferster and Skinner, 1957), responding more and more rapidly as the end of the interval approaches. Overall response rate on different FI schedules typically does not fit the equation very well. However, Schneider (1969) has argued for a two-state analysis of well-learned FI performance. In the first state, beginning immediately after reinforcement, response rate is very low and approximately constant over different FI values. At some variable time, on the average about two thirds of the time through the FI, there is an abrupt transition (or breakpoint) to a high and approximately constant response rate. Rate of responding in the second state is an increasing, negatively accelerated function of reinforcement frequency in that state, much like the function obtained for VI schedules. Schneider suggested that in the second state, the subject is on a VI schedule, the interreinforcement intervals being determined by the bird's breakpoint distribution, which is variable enough to approximate the range of intervals in ordinary VI schedules. Confirming this suggestion is our finding that Equation 2.8 provides an accurate account of second-state response rate in Schneider's experiment, though again with somewhat higher parameter values than those usually found for pigeons on VI schedules. With a k of 147 responses per minute and an R_e of 24 reinforcements per hour, the equation accounts for 92.2% of the variance in mean second-state response rate for Schneider's four pigeons.

Equation 2.8 can be applied to the rate of responding to one alternative in a choice procedure. Equation 2.8 in a two-choice procedure becomes:

$$B_1 = \frac{kR_1}{R_1 + R_2 + R_e}. \tag{2.9}$$

The extra term in the denominator, R_2, is the reinforcement from the other alternative, otherwise Equation 2.9 is the same as Equation 2.8. The data show that Equation 2.9 does, in fact, predict the raw frequencies of alternatives that obey the matching law when expressed as *relative* frequencies. For example, Catania (1963b) had pigeons pecking at two keys for food reinforcement. A description of the procedural details can be found in Catania's paper and in Herrnstein (1970). The proportions of responses to the alternatives obeyed Equation 2.1.

The fit of Equation 2.9 to the data is shown in Figure 2.3. The left-hand panel shows response rate at each alternative as a function of its own reinforcement frequency when the total rate of reinforcement frequency to both alternatives was held constant at 40 per hour. The right-hand panel shows response rate when the reinforcement for Schedule 2 was held constant at 20 per hour and that for Schedule 1 varied from 0 to 40 per hour. The straight line and two curves represent the plot of Equation 2.9 with k and R_e values as shown in the figure. The same parameter values were used for the functions in both panels. Only one data point deviates substantially from the plotted function, and the equation accounts for 91.0% of the variance in response rate in the left-hand panel and 90.3% of the variance in the right-hand panel.

Can the equation be extended to other parameters of reinforcement besides frequency and to other measures of response strength besides rate of key pecking by pigeons or lever pressing by rats? What

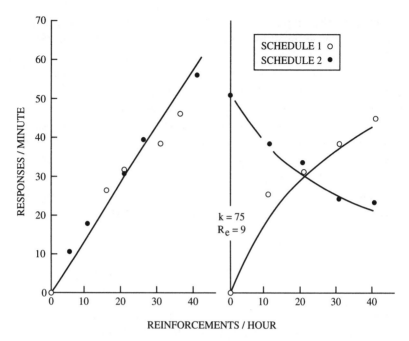

Figure 2.3   The absolute rate of responding as a function of the rate of reinforcement for each alternative in concurrent VI VI schedules. (For the left panel, the overall frequency of reinforcement was held constant at 40 reinforcements per hour, while varying complementarily for the two alternatives. Each point is here plotted above the reinforcement rate at which it was obtained. For the right panel, the frequency of reinforcement for Schedule 2 was held constant at 20 reinforcements per hour, while varying between 0 and 40 for Schedule 1. The points here are plotted above the reinforcement rate on Schedule 1 at the time it was obtained. The values of k and R_e were used for the smooth curves in both panels. Data taken from Catania (1963b).)

follows is a review of the literature on the relation between several measures of response strength (e.g., running speed in an alley, latency to respond in a discrete trial situation, as well as the rate of emission of repetitive responses like key pecking or lever pressing) and several parameters of both positive and negative reinforcement. For each study the data were either taken from tables or estimated from figures. We then executed the least squares fit to Equation 2.8 using individual data when available or group data otherwise. The analysis also estimated percentage of the variance in the dependent variable accounted for by the equation.

Although many experiments have examined the parameters of reinforcement since the crest of the Hullian wave in the 1940s, only about

40 are summarized here. To our knowledge, we have included all relevant experiments, but only a small fraction of this large literature can help us assess Equation 2.8. We are here interested in asymptotic performance under stable conditions of reinforcement. Many experiments in the Hullian tradition focused instead on acquisition or extinction curves, or on the effects of shifts in drive or incentive on acquisition or extinction. Where asymptotic performance has been measured, there are often only two values of the reinforcement variable, so that curve fitting to Equation 2.8 (with its two parameters) is meaningless. It may seem that even three data points are too few to test a two-parameter equation, and that studies with only three values of the reinforcement variable should therefore be excluded also. While we agree that such studies are only marginally useful taken individually, they have a cumulative impact that justifies including them.

Positive Reinforcement

Magnitude of Food Reinforcement

The two most extensive studies of magnitude of food reinforcement are those by Crespi (1942) and Zeaman (1949). In Crespi's study, five groups of 7–21 rats per group ran down a straight alley for different weights of dog chow, ranging from .02 to 5.12 g. Equation 2.8 describes the relation between the quantity of food and the mean running speed (100/time in seconds) for each group of rats remarkably well, accounting for 99.5% of the variance in running speed. In Zeaman's study, six groups of 8–10 rats per group ran down a short runway for amounts of cheese varying from .05 to 2.4 g. The response measured was the speed (100/latency in seconds) of getting out of the start box. Equation 2.8 accounts for about 82% of the data variance. Table 2.1 summarizes these and several other studies investigating the effects of magnitude of food reinforcement on response strength. In Davenport, Goodrich, and Hagguist's (1966) study, individual data were available for four macaque monkeys lever pressing for a varying number of pellets on a VI 1-minute schedule. Once again, Equation 2.8 provides a good fit to the data, accounting for 99.9%, 90.4%, 90.1%, and 96.6% of the variance in response rate for each monkey. For five of the remaining studies in Table 2.1 (Beier, 1958; Di Lollo, 1964; Hutt, 1954; Keesey and Kling, 1961, Experiment II; Logan, 1960, the 12-hour group), which show group data for rats and pigeons, Equation 2.8 accounts for at least 90% of the variance in the response measure in every case.

Table 2.1 Least squares fit of Equation 2.8 to the data from studies of magnitude of reinforcement

Study	k		R_e		% of variance accounted for
Crespi (1942)[a]	80	(running speed units)	.2	(g of chow)	99.5
Zeaman (1949)[b]	130	(100/latency in sec)	.3	(g of cheese)	82.1
Hutt (1954)[c]	12.4	(responses/min)	10.1	(mg of chow)	99.6
Beier (1958)[d]	68.0	(running speed units)	.67	(pellets)	99.4
Logan (1960)[e]					
12 hours deprived	45.1		.15		90.0
48 hours deprived	51.0	(running speed units)	.15	(pellets)	4.9
Keesey and Kling (1961)[f]					
Experiment I	118.6		.16	(peas)	12.2
Experiment II	67.0	(responses/min)	.45	(seeds)	99.3
Catania (1963a)[g]	64.6	(responses/min)	.02	(sec of grain access)	1.8
Di Lollo (1964)[h]	38.9	(running speed units)	1.0	(pellets)	99.7
Davenport, Goodrich, and Hagquist (1966)[i]					
Mean response rate of the four monkeys	21.7	(responses/min)	1.7	(pellets)	100.0
Individual monkeys					
S36	16.5		2.0		99.9
S49	30.0		3.0		90.4
S57	26.0		1.4		90.1
S60	17.0	(responses/min)	1.0	(pellets)	96.6

a. Running speed (100/time in seconds) in a straight alley for five different weights of dog chow between .02 and 5.12 g. Groups of 7–21 rats per value.

b. Speed (100/latency in seconds) to leave the start box of a short runway for six different weights of cheese between .05 and 2.4 g. Groups of 8–10 rats per value.

c. Rate of lever pressing on a periodic reinforcement 1-minute schedule for three different weights of rat chow between 3 and 5 mg. Groups of nine rats per value.

d. Running speed (100/time in seconds) in a straight alley for three numbers of pellets between 1 and 13 pellets. Groups of six rats per value.

e. Running speed (100/time in seconds) in a straight alley for four different numbers of food pellets between 1 and 12 pellets. Groups of six rats per reinforcement value and drive level.

f. Experiment I: Rate of key pecking on a VI 4-minute schedule for four different numbers of peas between one and four. Mean response rate of four pigeons, each exposed to all four values. Experiment II: Rate of key pecking on a VI 4-minute schedule for three different numbers of hemp seeds between two and eight seeds. Mean response rate of three pigeons, each exposed to all three values.

g. Rate of key pecking on a VI 2-minute schedule for three different durations of access to grain between 3 and 6 sec. Mean response rate of three pigeons, each exposed to all three values.

h. Running speed (100/time in seconds) in a straight alley for three different numbers of food pellets between 1 and 16 pellets. Groups of 16 rats per value.

i. Rate of lever pressing on a VI 1-minute schedule for three different numbers of sucrose pellets between one and nine pellets. Four macaque monkeys, each exposed to all three values.

We have found three exceptions to Equation 2.8 for magnitude of food reinforcement. Keesey and Kling (1961, Experiment I) studied four pigeons key pecking for different size chick peas (varying from four quarter peas to four whole peas) on a VI 4-minute schedule. Catania (1963a) studied two pigeons key pecking for three different durations of access to grain reinforcement (3, 4.5, and 6 sec) on a VI 2-minute schedule. Logan (1960, Experiment 55B) studied six rats per group, each group receiving a different number of pellets (1, 3, 6, and 12) for running down a straight alley. At 12 hours of food deprivation, the running speeds conformed to Equation 2.8, but not when deprivation was raised to 48 hours (see Table 2.1). The running speed at 48 hours of deprivation became a non-monotonic function of number of pellets, showing both a minimum and a maximum across the four values of the independent variable. In the two other deviant studies (see Catania, 1963a; Keesey and Kling, 1961, Experiment I—both in Table 2.1), the experimenters observed minimal changes in response rate with increases in reinforcement magnitude. Two factors could possibly account for the insensitivity of the key-peck rate to variations in reinforcement in these two studies. First, only a relatively narrow range of magnitudes was investigated, nowhere near the range studied by Crespi (1942) or the range of reinforcement in studies of frequency of food (Catania and Reynolds, 1968). Second, the pigeons were studied at high drive levels in both experiments (as were the rats in the disconfirming study by Logan). Equation 2.8 implies that when the underlying drive is high, R_e will be small relative to R_1 (Herrnstein, 1970). Where R_e is small relative to R_1, response rates should be close to k over a fairly wide range of R_1 values, since R_1 would approximate ΣR_i. It is thus noteworthy that in two of the cases in which Equation 2.8 accounts for little of the data variance, the data variance is minimal.

In brief, Table 2.1 indicates that Equation 2.8 can be extended to the relation between response strength and magnitude of food reinforcement for several responses and species, although with a few exceptions.

Brain Stimulation

The magnitude of reinforcement from intracranial brain stimulation varies with several parameters of stimulation—for example, the location, intensity, duration, or pulse frequency of the stimulation. But variations in response rate for brain stimulation on continuous reinforcement schedules do not provide an accurate measure of the effects of *magnitude* of reinforcement alone, since the *frequency* of brain stimulation is proportional to response rate on these schedules. At higher intensities or durations of stimulation, motor effects of the brain stimulation also tend to interfere with responding. Keesey (1962, 1964) therefore used a VI 16-sec schedule to investigate the effects of several parameters of electrical stimulation of the posterior hypothalamus on rate of lever pressing in rats. Table 2.2 shows that over a wide range of values, Equation 2.8 accurately depicts the relation between response rate and either the duration, intensity, or pulse frequency of the brain stimulation.

Table 2.2 Least squares fit of Equation 2.8 to the data from studies of reinforcement magnitude of brain stimulation

Study	*k*	R_e	% of variance accounted for
Keesey (1962)[a]			
Duration	15.1	.65 (msec)	94.1
Intensity	20.4	2.0 (mA)	92.9
Pulse frequency	25.0 (responses/min)	145 (pulses/sec)	96.9
Keesey (1964)[b]			
3.0 mA	21.0	.15	98.6
1.5 mA	14.9 (responses/min)	.10 (sec)	96.6
Gallistel (1969)[c]			
Individual rats			
LH 22	107.5	1.2	76.0
LH 30	111.5	5.0	84.2
LH 35	94.2	2.5	96.5
LH 54	115.2	1.2	78.5
DBB 35	106.0	3.8	96.2
DBB 38	110.5	11.9	90.2
DBB 44	110.0	4.0	92.9
PH 35	108.0	18.0	77.8
PH 40	102.0 (running speed units)	12.0 (pulses of brain stimulation)	95.4

a. Rate of lever pressing on a VI 16-sec schedule for six different durations (.28–2.65 msec), intensities (.5–4.0 mA), and pulse frequencies (23–138 pulses/sec) of hypothalamic brain stimulation. Mean response rates of 10 rats, each exposed to all six values of each parameter of brain stimulation.

b. Rate of lever pressing on a VI 16-sec schedule for five different durations (.25– 2.0 sec) of posterior hypothalamic brain stimulation at two different intensities (3.0 mA and 1.5 mA). Mean response rates of 10 rats, each exposed to all five durations at both intensities.

c. Running speed (100/time in seconds) in a straight alley for four to six different numbers of brain stimulation pulses (between 4 and 384 pulses/reinforcement). Nine rats, each exposed to several different stimulation values.

It is possible that the variance accounted for by the equation for intensity is below the other two (92.9% versus 94.1% and 96.9%) because the function for intensity passes through a maximum, whereas Equation 2.8 is monotonic. Keesey (1962) noted a slight nonmonotonicity for intensity at the upper end of the range examined, which might well have become more pronounced with still higher currents. But even with this complication, all five of Keesey's functions conform to Equation 2.8. In principle, however, we recognize that some reinforcement variables may indeed be nonmonotonic, in which case the fit to Equation 2.8 may require some rescaling of the independent variable.

Gallistel (1969) argued that lever pressing on a VI schedule of brain stimulation might not be a valid measure of the magnitude of reinforcement because response rate is also sensitive to the time since the last stimulation. To control for any variation in the aftereffects of the last brain stimulation, Gallistel gave all his subjects a series of 10 priming stimulations before they ran down a straight alley. Any differences in running speed should then be a function of the magnitude of the rewarding stimulation in the goal box. Nine rats, with electrodes implanted in three generally different rewarding areas of the brain, ran the straight alley for a varying number of .01-msec pulses of electrical stimulation. Table 2.2 shows that the least squares fit of Equation 2.8 accounts for a substantial proportion of the variance in running speed for each rat, even in some cases in which there was little variation in running speed, as was the case for LH35, LH22, and LH54 (LH meaning that the electrode tip was someplace in the lateral hypothalamus). The individual data have not been averaged here because of the variations in the electrode placement and because the various rats experienced different numbers of electrical pulses per reinforcement. It is nevertheless evident that Equation 2.8 accounts for the behavior reasonably well (i.e., better than 75%) in every instance and very well (i.e., better than 90%) for five out of the nine subjects.

Concentration of Sugar Reinforcement
Several experiments have investigated the effects of different sugar concentrations on response strength. Guttman (1954) studied seven concentrations of sucrose (between 2% and 32%, by weight) and the same seven concentrations of glucose with rats lever pressing on a VI 1-min schedule. The rats responded faster for sucrose than for glucose at each concentration, in keeping with the greater perceived sweetness of su-

crose for human subjects (Moskowitz, 1970), but Equation 2.8 accounts for the relation between response rate and concentration for both reinforcers, as Table 2.3 shows. The greater sweetness of sucrose appears to be well expressed by the smaller value fitted to R_e, reinforcement from extraneous sources. If the extraneous reinforcement is in fact constant in the two conditions, then measuring it in units of the more reinforcing sucrose should give a smaller value than measuring it in units of the less reinforcing glucose, as it does.

Table 2.3 summarizes the data from this and several other studies on concentration of sucrose. In Guttman's (1953) experiment two different conditions were studied. In one a different group of about 20 rats was exposed to each concentration; in the other, 20 rats were exposed to all four concentrations. In both cases the equation accounts for over 90% of the variance in mean response rate. In analyzing the data from rhesus monkeys obtained by Conrad and Sidman (1956), response rates for the 60% sucrose solution were excluded because the experimenters reported considerable satiation at this concentration. Two drive levels were used—48 and 72 hours of food deprivation. The main effect of deprivation appears to have been on R_e, with the higher drive yielding the lower value. By the same reasoning as we applied to the sucrose–glucose difference in Guttman's 1954 study, this is the expected result, further supporting the applicability of Equation 2.8.

A pair of more extensive studies by Schrier (1963, 1965), also with rhesus monkeys and a lever-press response, used considerably shorter sessions to avoid satiation effects. Over a range of concentrations from 10% to 50%, the equation fits both the average and individual data satisfactorily. For 10 of the 14 available individual functions, the variance accounted for was better than 90%, and for none did it fall below 70%. Schrier's three separate group averages were accounted for better than 95%. It is noteworthy that increasing the sample, as by averaging over individuals, increases the convergence on Equation 2.8.

Finally, Kraeling (1961) ran rats in a straight alley for sugar solution. Three concentrations and three durations of exposure to the sucrose reinforcer were varied factorially. There were 72 rats run altogether. Equation 2.8 was fitted to sucrose concentrations as the independent variable three times, once for each duration of exposure. The outcome, in Table 2.3, shows excellent confirmation, with high proportions of variance accounted for and the effects of increasing exposure largely confined to a decreasing value of R_e.

Table 2.3 Least squares fit of Equation 2.8 to the data from studies of sucrose and
glucose concentration

Study	k	R_e	% of variance accounted for
Guttman (1953)[a]			
Mean of 20 rats	8.5	4.2	92.4
Groups of rats	11.5 (responses/min)	11.5 (% sucrose)	95.6
Guttman (1954)[b]			
Sucrose	15.6	7.1 (% sucrose)	93.7
Glucose	16.1 (responses/min)	11.0 (% glucose)	98.7
Conrad and Sidman (1956)[c]			
48 hours deprivation	16.3	3.7	96.2
72 hours deprivation	17.6 (responses/min)	1.7 (% sucrose)	71.8
Kraeling (1961)[d]			
5 cc	89.4	2.4	100.0
25 cc	87.9	2.2	99.2
125 cc	88.7 (running speed units)	1.4 (% sucrose)	95.6
Schrier (1963)[e]	37.5 (responses/min)	6.1 (% sucrose)	95.4
Schrier (1965)[f]			
Mean response rate of the six monkeys exposed to both magnitudes			
.33 cc	88.0	16.1	95.9
.83 cc	67.2 (responses/min)	6.3 (% sucrose)	96.3
Individual data from the eight monkeys with .33 cc of sucrose			
Ruth	109.1	17.0	94.3
John	61.9	6.9	85.7
Ken	91.0	21.5	98.6
Allan	55.2	.8	70.1
Karen	87.5	9.5	95.2
Joan	131.2	52.4	92.6
Leo	105.9	40.1	98.6
Mae	6.6 (responses/min)	27.4 (% sucrose)	85.4

Table 2.3 (continued)

Study	k	R_e	% of variance accounted for
Individual data from the six monkeys with .83 cc of sucrose			
Ruth	82.4	5.5	84.9
John	77.2	6.0	97.0
Ken	86.5	20.0	97.7
Allan	61.9	7.5	90.8
Karen	68.2	11.2	93.3
Joan	45.8 (responses/min)	65.0 (% sucrose)	98.3

a. Rate of lever pressing on a periodic reinforcement 1-minute schedule for four different concentrations of sucrose (4%–32%). Mean response rates of 20 rats, each exposed to all four values; and mean response rates of groups of 20 rats per value.

b. Rate of lever pressing on a VI 1-minute schedule for seven different concentrations of sucrose and glucose (2%–32%). Mean response rates of eight rats, each exposed to all seven values of both reinforcers.

c. Rate of lever pressing on a VI 37-sec schedule for five different concentrations of sucrose (2.3%–30%) at two drive levels (48 and 72 hours of deprivation). Mean of three monkeys, each exposed to all five concentrations at both drive levels.

d. Running speed (100/time in seconds) in a straight alley for three different concentrations (2.4%–9.1%) and three different magnitudes (5 cc–125 cc) of sucrose. Groups of nine rats for each concentration and magnitude.

e. Rate of lever pressing on a VI 1-minute schedule for five different concentrations of sucrose (10%–50%). Mean response rates of four monkeys, each exposed to all five concentrations.

f. Rate of lever pressing on a VI 30-sec schedule for five different concentrations (10%–50%) and two different magnitudes (.33 cc and .83 cc) of sucrose. Response rates of eight and six monkeys, each exposed to all five concentrations; six of them exposed to both magnitudes.

Immediacy of Positive Reinforcement (1/Delay)

The most comprehensive data on delay of reinforcement in a single response situation we owe to Pierce, Hanford, and Zimmerman (1972). Four rats responding on a VI 1-min schedule for food reinforcement experienced delays of reinforcement varying from .5 to 100 sec between the reinforced lever press and the operation of the feeder. During the delay a cue light was illuminated and responding had no programmed consequences. Equation 2.8 again provides an accurate description of Pierce et al.'s results, with immediacy of reinforcement (1/delay) as R_1. It accounts for 96.1% of the variance in mean response rate and for 80.8%, 98.3%, 78.9%, and 97.4% of the variance for the four individual rats. The same rats were studied in a second condition, in which the lever was retracted during the delay of reinforcement. The equation fits the data from this condition about as well, accounting for 95.0% of the variance in mean response rate and for 97.8%, 98.6%, 94.9%, and 92.9% of the variance for the four individual rats.

Table 2.4 gives the least squares fit of Equation 2.8 to several other studies on delay of reinforcement as well as that of Pierce et al. (1972). The Perin (1943) experiment is noteworthy in that it measured latency to lever press in a discrete trial procedure as a function of delay of reinforcement, rather than frequency of lever pressing, which is the more common measure of response strength. The equation nevertheless accounts for much of the relation between speed of responding (100/latency in seconds) and immediacy of food presentation (1/delay). Perin used five conditions of reinforcement delay—0, 2, 5, 10, and 30 sec—but Table 2.4 reports the results of fitting Equation 2.8 to only three of them—2, 5, and 10 sec. The 30-sec group has been left out because Perin reported that it failed to reach an asymptotic latency of response in the relatively brief span of the experiment, and also because some of the rats in the group may have been extinguishing in lever pressing.

The omission of the 0-delay group raises a more interesting point. If we take the reciprocal of delay to be the proper dimension of reinforcement, then the 0-delay group is receiving an incalculably large reinforcement (i.e., 1/0 sec). Equation 2.8 under those conditions predicts that response speed should approach the value of k (i.e., 54 in units of speed), for as R_1 goes to infinity, the value of $R_1/(R_1 + R_e)$ goes to 1.0. The observed speed at 0 delay was, however, 47.0 (100/latency), which may be close enough by usual standards of curve fitting. We are uneasy with this treatment of the data because it presupposes that the 0-delay group actually does receive its reinforcement instantaneously, which is obviously contrary to fact. The rat presses the lever and it is clearly a moment or two before the food is taken. The duration of that gap between response and reinforcement should be plugged into our least squares procedure. But because the actual reinforcement delay is unknown, we have instead fitted Equation 2.8 to the more comparable three other values—2, 5, and 10 sec—as shown in Table 2.4. With the obtained parameters, we can then estimate the actual delay of reinforcement for the nominal 0-delay condition.

Fitting $k = 54.0$, $R_e = .29$, and the obtained response speed for B_1 (i.e., 47.0) into Equation 2.8, the solved value for R_1 is 1.95 when the

Table 2.4   Least squares fit of Equation 2.8 to the data from studies of immediacy of positive reinforcement

Study	k	R_e (1/delay in seconds)	% of variance accounted for
Perin (1943)[a]	54.0 (100/latency in sec)	.29	93.2
Logan (1960, Exper. 55D)[b]			
High drive	64.9	.016	99.1
Low drive	64.5 (running speed units)	.075	92.8
Silver and Pierce (1969)[c]	9.8 (responses/min)	.02	92.8
Pierce, Hanford, and Zimmerman (1972)[d]			
Cue light (mean)	21.4	.04	96.1
R1	21.5	.03	80.8
R2	23.5	.15	98.3
R3	22.8	.02	78.9
R4	20.9 (responses/min)	.07	97.4
Lever retracted (mean)	39.4	.09	95.0
R1	219.1	.63	97.8
R2	67.5	.80	98.6
R3	102.3	.23	94.9
R4	26.6 (responses/min)	.04	92.9

a. Speed of responding (100/latency in seconds to lever press) in a discrete trial situation with three different immediacies of reinforcement (2–10 sec delay). Groups of 25 rats per delay value.

b. Running speed (100/time in seconds) in a straight alley for five different immediacies of reinforcement (1–30 sec delay). Groups of 10 rats per delay value, each run at high and low drive.

c. Rate of lever pressing on a VI 1-minute schedule with five different immediacies of reinforcement (10–160 sec delay). Mean of six rats, each exposed to all five delay values.

d. Rate of lever pressing on a VI 1-minute schedule with five different immediacies of reinforcement (.5–100 sec delay), the delay of reinforcement being signaled by a cue light. Also three immediacies of reinforcement (10–100 sec delay) with the lever retracted during the delay. Mean response rates of four rats, each exposed to all of the delay values in both delay conditions.

nominal delay is 0. The solved value, 1.95, is the immediacy of reinforce-ment, whose reciprocal is the actual delay (i.e., .51 seconds). This says that the actual delay for the nominal 0-delay condition was about half a second, a conclusion that squares quite well with our general impression of such lever-pressing procedures.

The two other sources (Logan, 1960; Silver and Pierce, 1969) sum-marized in Table 2.4 also conform substantially to Equation 2.8. Both experiments used rats as subjects and food as reinforcement, but Logan's data are for speed in a runway and Silver and Pierce's are for rate of lever pressing. Logan had high and low drive conditions, which pro-duced faster and slower running, respectively. As before, the effect of drive appears to be well handled by an adjustment in the value of R_e. In each case, variance in group data is better than 90% accounted for by Equation 2.8.

Negative Reinforcement

Frequency of Negative Reinforcement

De Villiers (1974) demonstrated that Equation 2.8 can be extended to VI avoidance schedules, with shock-frequency reduction (shocks avoided per minute) as the reinforcer for avoidance (Herrnstein, 1969; Herrn-stein and Hineline, 1966). In de Villiers' experiment, lever-press re-sponses by rats cancelled the delivery of shocks scheduled at variable intervals. If no lever press was made, all of the scheduled shocks were presented. The first response made after a scheduled shock, whether or not that shock had been presented, prevented the delivery of only the next scheduled shock. Extra responses between two scheduled shocks did not allow the rats to avoid further shocks. All the scheduled shocks could therefore be avoided if the rat responded at least once within every intershock interval, but the durations of the intervals varied un-predictably. On this VI avoidance schedule both the shock rate actually received and the shock-frequency reduction (scheduled shock rate minus received shock rate) can be measured, and response rate is not con-strained by any fixed temporal relations between responses and shocks, as it is on the free operant avoidance schedules usually studied (Sidman, 1953, 1966).

Four rats were exposed to an ascending and descending series of VI avoidance schedules ranging from a VI 15-sec (4 programmed shocks per minute) to a VI 75-sec (.8 programmed shocks per minute). A wide range of response rates was produced by each rat on the different schedules.

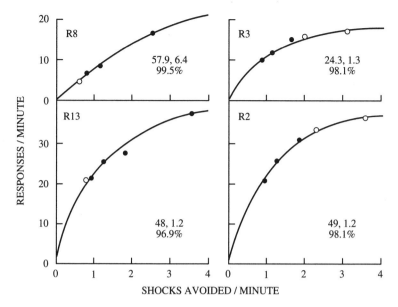

Figure 2.4   Rate of responding as a function of shock-frequency reduction for de Villiers' (1974) four rats responding on single VI avoidance schedules. (The least squares fit of Equation 2.8 to the data is plotted for each rat. The k and R_e values and the percentage of the variance in response rate accounted for by the equation are also shown. Filled circles represent the mean response rate of two determinations for a given VI value; open circles represent single determinations.)

Response rates are plotted against shock-frequency reduction (shocks *avoided* per minute) for each rat in Figure 2.4. The least squares fit of Equation 2.8 is shown for each animal. Where there were two determinations of response rate for a given VI value, the mean of the two was used in calculating the best fit of the equation. Equation 2.8 accounts for over 95% of the variance in response rate for each rat: 99.5% for R8, 96.9% for R13, 98.1% for R3, and 98.1% for R2.

Magnitude of Negative Reinforcement

Several experiments have investigated the effects of different amounts of voltage reduction between the alley and goal box on a rat's running speed in a straight runway. Three of these studies (Campbell and Kraeling, 1953; Bower, Fowler, and Trapold, 1959; Seward, Shea, Uyeda, and Raskin, 1960) are summarized in Table 2.5. All of them used different groups of rats for each voltage reduction value. The rat would run down a straight alley with an electrified floor to reach an end box whose floor

Table 2.5 Least squares fit of Equation 2.8 to the data from studies of magnitude of negative reinforcement

Study	k	R_e	% of variance accounted for
Campbell and Kraeling (1953)[a]			
400-V alley	228.0	701.0	92.9
300-V alley	106.0 (running speed units)	125.0 (V)	98.9
Bower, Fowler, and Trapold (1959)[b]	185.0 (running speed units)	338.0 (V)	99.6
Seward, Shea, Uyeda, and Raskin (1960)[c]			
315-V alley	161.0 (running speed units)	52.0 (V)	97.7
255-V alley	146.0 (running speed units)	35.0 (V)	86.0
Woods, Davidson, and Peters (1964)[d]	163.0 (swimming speed units)	18.0 (°C)	93.6
Woods and Holland (1966)[e]			
25°C tank water	108.0	1.3	85.9
15°C tank water	114.0 (swimming speed units)	1.4 (°C)	87.5
Dinsmoor and Hughes (1956)[f]			
.2 mA	45.6	89.4	99.5
.4 mA	24.3 (100/latency in sec)	6.7 (sec time-out)	74.1
Harrison and Abelson (1959)[g]	3.1 (responses/min)	.45 (sec time-out)	91.4

a. Running speed (100/time in seconds) in a straight alley for four different reductions in voltage (100–400 V) with a 400-V alley and for three different voltage reductions (100–300 V) with a 300-V alley. Groups of seven rats for each voltage reduction and each alley intensity.

b. Running speed (100/time in seconds) in a straight alley (250-V alley intensity) for three different reductions in voltage (50–200 V). Groups of five rats for each voltage reduction value.

c. Running speed (100/time in seconds) in a straight alley for three different voltage reductions (65–315 V) with two different alley intensities (315 and 255 V). Groups of nine rats for each voltage reduction and alley intensity.

d. Swimming speed (100/time in seconds) in a cold water tank (15°C water) for three different increases in water temperature (5°–25°C). Groups of 10 rats per temperature increase.

e. Swimming speed (100/time in seconds) in a cold water tank for three different increases in temperature (4°C–16°C) with two different water tank temperatures (15° and 25°C). Groups of 16 rats for each temperature increase and water tank temperature.

f. Response speed (100/latency in seconds to lever press) for four different durations of time-out (5–40 sec) from .2 mA and .4 mA shock. Groups of five rats per value of time-out and intensity.

g. Rate of lever pressing for four different durations of time-out (2–20 sec) from loud noise (116 dB.). One rat exposed to all four time-out durations.

was electrified by a smaller voltage. The several experiments employed a variety of runway and end-box voltages. In fitting Equation 2.8, we took running speed as the response measure and the reduction in voltage (expressed as a positive number) as the reinforcer, which determines the proper interpretation of k and R_e. It may seem paradoxical to express extraneous reinforcement, R_e as a voltage, but it is really in units of voltage *reduction*. In Bower et al. (1959), for example, the extraneous reinforcement equaled a reduction of 338 V. The value of R_e seems much smaller in Seward et al. (1960), but this may be partly or entirely artifactual. Seward et al. used a 150,000 Ω series resistance in their electrical circuit, whereas the other two experiments used 250,000 Ω. Each volt in Seward et al. therefore represents a substantially larger shock than that in the other two. For all three experiments, Equation 2.8 accounts for at least 85% of the data variance, and usually for more than 95%.

Woods and his co-workers (Woods, Davidson, and Peters, 1964; Woods and Holland, 1966) used an interesting variant of the runway escape procedure. Their rats swam down a straight alley filled with cold water in order to be placed in a goal box containing warmer water. The independent variable was the increase in water temperature between the alley and the goal box; the dependent variable was swimming speed (100/time). Equation 2.8 accounts for these data very well, although it is not clear why the parameter values in the two studies differ as much as they do. From the published account it is likely that the rats in the second study were younger, therefore more subject to the temperature reinforcement. If so, the smaller value for R_e would be expected.

Dinsmoor and Hughes (1956) and Harrison and Abelson (1959) varied the duration of time-out from aversive stimulation contingent on a lever-press escape response in rats. Dinsmoor and Hughes measured latency to lever press in a discrete trial escape procedure and observed a monotonically increasing function relating speed of response (100/latency) and duration of time-out. The equation accounts for 99.5% of the variance in speed of response for the rats that were escaping from a .2-mA shock. The variance in speed of lever pressing for rats escaping from the stronger shock, .4 mA, was only about 75% accounted for. This may be yet one more instance of higher drives allowing so little variance in behavior that curve fitting begins to become uninformative.

Harrison and Abelson measured response rate on VI escape from a loud noise. Although they found strong order effects and exposed only one rat to complete ascending and descending orders, we have applied Equation 2.8 to that one rat's behavior. If response rates are averaged across the several determinations for each time-out duration, over 90% of the variance is accounted for by Equation 2.8.

Immediacy of Negative Reinforcement

Fowler and Trapold (1962) conducted a thorough study of delay of escape in a straight runway. Delay of voltage reduction in the goal box varied over five values between 1 and 16 sec for groups of five rats each. The group with 0 delay is omitted from the analysis, as before. The observed relation between running speed in the runway (100/time) and immediacy of escape (1/delay) in the goal box conforms to Equation 2.8, 92.6% of the variance in running speed being accounted for by the equation with a k of 94.0 (running speed units) and an R_e of .06 (1/delay units). These results, along with other data from experiments varying delay of negative reinforcement, appear in Table 2.6.

Tarpy (1969) caused rats to escape from an electric shock by pressing either of two levers. The primary variable was the delay between a lever press and the shutting off of the shock, which was typically, though not always, set at different values for the two levers. The trials were spaced at 1-minute intervals (not including the duration of the delay interval). The experiment's main purpose was to examine preference as a function of delay, but we have used response latency to test Equation 2.8. Table 2.6 shows the outcome, based only on the subjects that had equal delays for both levers, since those were the only subjects for which latencies were given. About 89% of the variance in response speed (100/latency) for five delays (1–16 sec, again excluding the 0-delay condition) is accounted for. Tarpy and Koster (1970) then varied delay of escape from electric shock by rats lever pressing in a simple, one-response procedure. Using the three non-zero values of delay, about 95% of the data variance in response speed was accounted for by Equation 2.8.

Leeming and Robinson (1973) also varied delay of escape by rats from electric shock, but the escape response was running from one compartment in a shuttle box to the other. Equation 2.8 accounted for about 86% of the variance in response speed (100/latency in seconds) for the subjects receiving the five nonzero delays of shock termination. In this experiment both the measure of response latency and the timing of delays included the experimenters' reaction times, since the apparatus was not fully automatic. This was doubtless an additional source of variance.

Finally, Table 2.6 also summarizes Moffat and Koch's (1973) results on human subjects listening to a Bill Cosby comedy recording. Occasionally the recording would stop and the subject could start it again

Table 2.6   Least squares fit of Equation 2.8 to the data from studies of
immediacy of negative reinforcement

Study	k	R_e (1/delay in seconds)	% of variance accounted for
Fowler and Trapold (1962)[a]	94.0 (running speed units)	.06	92.6
Tarpy (1969)[b]	98.1 (100/latency in sec)	.33	89.5
Tarpy and Koster (1970)[c]	45.5 (100/latency in sec)	.40	94.6
Leeming and Robinson (1973)[d]	39.4 (100/latency in sec)	.11	86.0
Moffat and Koch (1973)[e]	304.1 (1/latency in .01 sec)	2.0	96.6

a. Running speed (100/time in seconds) in a straight alley (240-V alley intensity) with shock offset in the goal box delayed for five different delays between 1 and 16 sec. Groups of five rats per delay value.

b. Speed (100/latency in seconds) of lever-press escape response on either of two levers in a discrete trial situation for five different immediacies of offset of a 200-V shock (1–16-sec delay). Groups of 10 rats per delay value.

c. Speed (100/latency in seconds) of lever-press escape response in a discrete trial situation for three different immediacies of offset of a 200-V shock (1.5–6-sec delay). Groups of 10 rats per delay value.

d. Speed (100/latency in seconds) of escape in a shuttle box for five different immediacies of offset of a 420-V shock (1–16-sec delay). Groups of 10 rats per delay value.

e. Speed (1/latency in .01 seconds) of panel depression to escape time-out from a comedy record for three different immediacies of reinstatement of the record (3–9-sec delay). Groups of 10 human subjects per delay value.

by depressing a panel. The primary variable was the delay between the panel response and the onset of the recording. For the three nonzero delays, Equation 2.8 accounts for about 97% of the variance in speed (1/latency) of the panel depression.

Summary of Findings

The generality of Equation 2.8 is apparent from this survey. The behavior of rats, pigeons, monkeys, and, in the one case we have found, people is fairly well accounted for, whether the behavior was lever press-

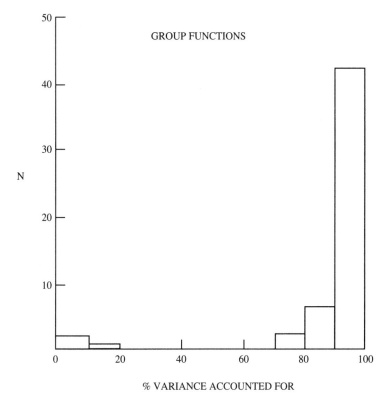

Figure 2.5   Equation 2.8 was applied to 53 studies using group data. (The variances accounted for are tallied in class intervals of 10%.)

ing, key pecking, running speed, or response latency, considering the diversity in experimental settings from one laboratory to the other over the past 30 years. The reinforcers were as different as sweetened water, food, escape from shock or loud noise or cold water, electrical stimulation of a variety of brain loci, and turning a comedy recording back on. Figure 2.5 presents a histogram of the variance accounted for by the least squares fit of Equation 2.8 to the 53 studies involving group data reviewed in this paper. A large majority of cases fall above 90% and only 3 of the 53 fall below 70%. Much the same picture is conveyed in Figure 2.6, which includes the 49 cases in which individual data were used here to test Equation 2.8. While the peak in Figure 2.6 may be slightly blunter than that in Figure 2.5, the individual data contain no case of less than 70% of the variance accounted for by Equation 2.8. The literature appears to contain no evidence for a substan-

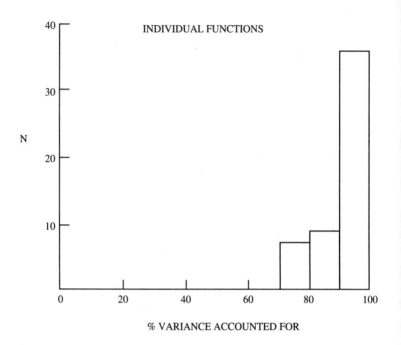

Figure 2.6   Equation 2.8 was applied to 49 individual functions. (The variances accounted for are tallied in class intervals of 10%.)

tially different equation from Equation 2.8. Where the equation fails to account for most of the data variance, the data variance is negligible in every case but one. The one exception (Logan, 1960, Experiment 55B) involves an experiment in which supposedly asymptotic performance was a nonmonotonic function of reinforcement magnitude. Other experiments on magnitude, including one by Logan, found monotonic functions, all accounted for by Equation 2.8. This equation therefore provides a powerful yet simple framework for the quantification of the relation between response strength and both positive and negative reinforcement.

While it is clear that Equation 2.8 fits the data satisfactorily, it remains to be shown that there are not equal or better alternatives. A number of the studies summarized had only three data points, as we noted earlier. Other downwardly concave functions with a couple of parameters could no doubt have handled those cases as well, or nearly as well, as Equation 2.8. We have checked the two more obvious alternatives—exponential and power functions—and have found them mostly inferior to Equation 2.8 in fitting the data from studies in which there were *more* than three data points.

Table 2.7 compares the variances accounted for by least squares fits to Equation 2.8 to the exponential,

$$B_1 = k(1 - e^{-R_1/R_e}), \tag{2.10}$$

and to the power function,

$$B_1 = kR_1^{R_e}. \tag{2.11}$$

Each of these three equations involves two free parameters, though not necessarily with the same empirical meanings. Also, they all can take downwardly concave shapes, which is an obvious requirement for any quantitative theory that presumes to account for the data. Table 2.7 compares only those studies shown in Tables 2.1–2.6 that used five or more values of the reinforcement variable (for procedural specifics, see Tables 2.1–2.6). Although all three equations handle most of the data variance in these 23 studies, Equation 2.8 is, on the average, a significantly better fit than is Equation 2.10 or 2.11. Nevertheless, Equation 2.11 appears to be best for the three studies on the immediacy of negative reinforcement. Two of these three studies used response latency to measure behavior, as did Zeaman (1949), which also showed a marked deviation in favor of the power function (Equation 2.11). We can offer no explanation of the apparent superiority of power functions for latency measures.

The case for Equation 2.8 as the principle of response strength rests only partly on the goodness of fit to existing data. It also rests on the relation between it and the matching principle for choice, as noted at the beginning of this paper. Equation 2.8 handles not only simple frequencies of behavior but it also implies a well-substantiated prediction about the way behavior is distributed across alternatives. While quantity of behavior intuitively seems to be a simpler psychological dimension than does choice, in our analysis the reverse is the case. The matching law for symmetric simple choice (Equation 2.1 above) involves no parameters, only the measured variables of response and reinforcement. Equation 2.8, for quantity of behavior, involves the same variables plus two parameters, k and R_e. To extend the analysis from simple choice to absolute response frequency is thus an advance in both scope and complexity.

Perhaps the most surprising feature of the close fit between data and theory is that it has not required elaborate or ad hoc transformations of

Table 2.7 Percentage of variance accounted for by three comparable equations

Study	Equations				
	8	10	11	8–10[a]	8–11[b]
Positive reinforcement					
Frequency					
Catania and Reynolds (1968)	91.3	81.7	90.7	+9.6	+.6
Magnitude					
Crespi (1942)	99.5	97.1	91.1	+2.4	+8.4
Zeaman (1949)	82.1	74.9	95.7	+7.2	−13.6
Brain stimulation					
Keesey (1962)					
Frequency	96.9	96.8	93.7	+.1	+3.2
Duration	94.1	94.1	91.4	0	+2.7
Intensity	92.9	92.9	91.3	0	+1.6
Keesey (1964)					
1.5 mA	96.6	86.0	95.9	+10.6	+.7
3.0 mA	98.6	91.4	95.7	+7.2	+2.9
Sugar concentration					
Guttman (1954)					
Sucrose	93.7	93.7	89.6	0	+4.1
Glucose	98.7	99.3	94.4	−.6	+4.3
Conrad and Sidman (1956)					
48 hr	96.2	94.8	90.6	+1.4	+5.6
72 hr	71.8	70.9	60.8	+.9	+11.0
Schrier (1963)	95.4	99.6	85.4	−4.2	+10.0
Schrier (1965)					
.33 cc	95.9	90.1	98.5	+5.8	−2.6
.83 cc	96.3	86.5	96.6	+9.8	−.3
Immediacy of reinforcement					
Logan (1960)					
High drive	99.1	84.6	87.2	+14.5	+11.9
Low drive	92.8	91.0	76.6	+1.8	+16.2
Silver and Pierce (1969)	92.8	96.6	81.1	−3.8	+11.7
Pierce, Hanford, and Zimmerman (1972)	96.1	98.1	75.1	−2.0	+21.0

Table 2.7 (continued)

	Equations				
Study	8	10	11	8–10[a]	8–11[b]
Negative reinforcement					
Frequency					
de Villiers (1974)	98.2	97.9	97.5	+.3	+.7
Immediacy of reinforcement					
Fowler and Trapold (1962)	92.6	75.4	99.0	+17.2	−6.4
Tarpy (1969)	89.5	86.8	93.6	+2.7	−4.1
Leeming and Robinson (1973)	86.0	73.2	96.5	+12.8	−10.5

Note. Equation 2.8 is $B_1 = \frac{kR_1}{R_1 + R_e}$; Equation 2.10 is $B_1 = k(1 - e^{-R_1/R_e})$; Equation 2.11 is $B_1 = kR_1^{R_e}$.

a. Mean difference of Equation 2.8 − Equation 2.10 = +4.0%. Wilcoxon matched pairs 8 versus 10 ($p < .01$).

b. Mean difference of Equation 2.8 − Equation 2.11 = +3.5%. Wilcoxon matched pairs 8 versus 11 ($p < .05$).

the variables. For example, studies of sugar reinforcement (see Table 2.3) or brain stimulation (see Table 2.2) are well accounted for by Equation 2.8 taking the independent variable as measured on ordinary physical scales, such as percent concentration by weight or electrical pulses per second. In contrast, psychophysics teaches us that many dimensions of experience are not accounted foɪ by physical scales unless the physical scale is somehow transformed—the specification of which is the essence of psychophysics (Marks, 1974). The very simplicity of the independent variables here is consequently a riddle, for which we offer four nonexclusive, partial solutions.

First, several of the independent variables have been shown to produce matching of relative response to relative reinforcement (see Equation 2.1). Reinforcement rate, magnitude, and delay fall into this category. Variables that obey Equation 2.1 should also obey Equation 2.8, for the very reasons given at the beginning of this chapter. The fit of Equation 2.8 to data on rate, magnitude, and delay of reinforcement is in that sense only a further confirmation of the matching law for choice.

Second, it is possible that some of the simplicity reflects the relatively small ranges of the independent variable used in studies of performance. The range from one to nine pellets of sugar or .28 to 2.7 msec of brain stimulation is not quite an order of magnitude, whereas proper psychophysical experiments use several orders of magnitude. Even if the true reinforcement variable were not amount of sugar or duration of stimulation but curvilinear functions of them, over the typical range of behavioral experiments it may appear linear. Perhaps Equation 2.8 usually accounts for something less than 100% of the data variance partly because we have failed to rescale the physical variables to make them more valid psychologically. When behavioral experiments begin to use the broad spans of stimuli found in psychophysical research, Equation 2.8 may be the best way to find out what is, in fact, the correct way to specify the reinforcement variable.

Third, the people who experiment on reinforcement variables draw on intuition to pick their variables. For example, they vary the number, which is to say, the *mass* of pellets of chow rather than the *diameter* of each pellet. Why? If Equation 2.8 accounts for the effect of the mass of pellets, it *cannot*, mathematically, account for the effect of diameter, since the one variable is a nonlinear function of the other. For each dimension varied in the experiments reviewed here, there are an unlimited number of nonlinearly related physical dimensions that would have been only poorly accounted for by Equation 2.8. It is no accident that the right variable gets varied in experiment after experiment. We know the dimensions of reinforcement to at least a first approximation because we are ourselves reinforced by them or their analogs. We know that, for example, concentration measures what is reinforcing about a sugar solution better than, say, viscosity, or that what mainly counts is the amount of chow not its length or shape. It is important to make this intuitive selection of variable explicit, for it is bound to be wrong at times. Viscosity of solutions, shape of pellets, and so on, may have independent status as parameters of reinforcement (positive or negative), even if the status is minor, compared to the biologically obvious ones usually experimented on. Here, as before, the result will at times be a certain amount of variance in the data unaccounted for.

Finally, the reinforcement variables we have examined have been largely confined to the parameters of incentive rather than of drives. We assess Equation 2.8 with amount of food reinforcement as an independent variable but not with hours of food deprivation or percentage of ad libitum body weight. The reason is not a lack of data, for many experiments look at performance as a function of drive variables like deprivation or body weight. These variables seem to be too indirect a measurement of the reinforcement dimensions to which they doubtless relate. Some function of hours of deprivation or percentage of ad lib weight may be expressible as the food reinforcement variable in Equation 2.8, but whatever the function is for food reinforcement for a given species, we can see no argument for supposing it to have any generality across drives. Equation 2.8 could possibly be used to calibrate drive operations, estimating, for example, how many hours of food deprivation are equivalent, in terms of the reinforcing power of a unit of food, to a given percentage of ad lib weight. But the data needed for such calibrations do not seem to be available at this time.

There are some data, however, for evaluating another point of contact between drive variables and Equation 2.8. In a recent paper Herrnstein (1974) argued that a formal implication of the equation is that for any given response, the parameter k must remain constant across different qualities or quantities of reinforcement or changes in the animal's drive level, so long as the response topography does not change (see Equations 2.3–2.6). The k is just a measure of behavior, such as responses per minute. It represents the amount of behavior the observed response would show if there were no reinforcement for competing responses. When there are other sources of reinforcement, k measures the total frequency of all responses in units commensurate with the measure of the response being studied. The k is therefore simply "the modulus for measuring behavior" (Herrnstein, 1974), and the sole influence on its size is the chosen response form itself.

Several of the studies summarized here allow tests of constancy of k across different qualities and quantities of reinforcement or across different levels of drive. When Equation 2.8 was fitted to Guttman's data on the same eight rats lever pressing for either sucrose or glucose (Table 2.3), practically the same k value was derived for the two reinforcers (15.6 versus 16.1 responses per minute). The reinforcement from other sources (R_e) was smaller for the sucrose function (7.1% versus 11.0%), in keeping with rats' preference for sucrose, the sweeter solution. Kraeling (1961) ran different groups of rats in a runway to sugar solutions of varying concentration. Three separate functions between running speed and concentration were found, one for each of three different durations of exposure to the sugar water at the end of the runway (Table 2.3). The question is whether the same k was appropriate to each function, and the answer is that the three ks were 89.4, 87.9, and 88.7 running speed units, which is impressive confirmation considering that different rats ran in the different conditions. Logan (1960) varied immediacy of reinforcement under high and low drives for food. The ks fitted to his data (Table 2.4) are 64.9 and 64.5, again impressive confirmation of the constancy of k. Seward et al. (1960) ran rats in runways to escape from higher to lower voltages of electric shock. In our analysis (Table 2.5), we took voltage reduction as the reinforcement variable and alley voltage as the drive variable. The two alley voltages yielded ks of 161 and 146 running speed units, which is a discrepancy of about 10%. In the comparable experiment on water temperature (Woods and Holland, 1966; see Table 2.5), the rats swam from cold water to warmer water. The ks for the two alley temperatures were 108 and 114 swimming speed units, which is better agreement than Seward et al. found for the electric shock version of the experiment.

Although the foregoing supports the hypothesis that k in Equation 2.8 remains roughly constant as reinforcement and drive parameters change, some of the data analyzed do not. In both of Keesey's experiments (1962, 1964; see Table 2.2) on rats lever pressing for electrical stimulation of the brain, there are substantial variations in the value of k. The five obtained values were 15.1, 20.4, 25.0, 21.0, and 14.9 responses per minute—in the same neighborhood but obviously not the same. In Schrier's experiment (1965; see Table 2.3) on monkeys lever pressing for sugar water, the two values of k were 88 and 67.2 responses per minute for two different reinforcement quantities. Finally, the largest discrepancy was in Campbell and Kraeling's (1953; see Table 2.5) experiment on rats running from more to less intense shock, in which the two ks were 228 and 106 running speed units. If the assumption of constant response topography was not violated in these disconfirming studies, then some modification of our analysis of response strength will be in order. In the meantime, the quantitative precision of Equation 2.8 encourages us to believe that we are at least near a law of response strength for reinforced behavior. We can conclude only that further experiments must investigate the range of conditions under which k remains or does not remain constant.

3

Derivatives of Matching

The matching law arose empirically in experiments on simple choice. In numerous studies, pigeons pecking keys or rats pressing levers to earn food or other reinforcers distributed responses in approximately the same ratios as the frequencies of reinforcement (de Villiers, 1977). If, say, 10% of the reinforcements came for pecking the left key, then about 10% of the pecks were at the left key, even though there might have been scores of pecks for each reinforcement. Something about the dynamics of reinforcement constrains subjects to equalize the amount of behavior per reinforcement across choices.

As an empirical generalization, matching has been unfolding for more than 15 years (see Chapter 1) and is not yet a complete story. Although few now question its approximate accuracy, questions continue to be raised about matching as an exact principle (Myers and Myers, 1977) and about its relationship to maximization of reinforcement. Shimp (1966, 1969, 1975) and others (e.g., Rachlin, Green, Kagel, and Battalio, 1976; Staddon and Motheral, 1978) have tried to identify a maximization process to explain matching. In some writings (Mackintosh, 1974; Shimp, 1975; Silberberg, Hamilton, Ziriax, and Casey, 1978) matching is conceived as a molar approximation to more fundamental and molecular principles of behavior. In contrast, Rachlin

Originally published in *Psychological Review*, 1979, *86*, 486–495. Copyright © 1979 by the American Psychological Association. Reprinted with permission. Preparation of this essay was supported by Grant MH-15494 from the National Institute of Mental Health to Harvard University. Thanks are due to J. E. Mazur and R. Hastie for sharing their reanalyses of learning curves, which provided much of the impetus for the section on acquisition. R. D. Luce, H. Rachlin, and several anonymous reviewers commented helpfully on earlier drafts, for which the author is grateful. A version of this essay was read at the 1976 meeting of the American Psychological Association.

(1971) has said that the matching law is a tautology, true by definition hence not subject to empirical test. Killeen (1972), responding to Rachlin, agreed that the matching law is an implicit definition of utility (i.e., reinforcing value), but added that empirical findings promote this particular definition rather than the many others that might have been advanced.

The main purpose of this article is not to review the literature but to relate matching to the molecular level of analysis. The result, as should become clear, is an extension of Rachlin's (1971) and Killeen's (1972) exchange on matching as a definition of reinforcement. The derivation given below overlaps Staddon's (1977), but the theoretical context and some of the conclusions are different. Gibbon's (1977) theory of scalar expectancy may also provide a molecular account of matching but he has not yet published on its implications for the aperiodic schedules of reinforcement most often used. Implicitly, all such attempts, including this one, are predicated on the assumption that molar matching must be at least reconciled with, if not derived from, molecular processes.

The matching law says that responding is proportional to relative reinforcement, as follows:

$$B = \frac{kR}{\Sigma R}. \tag{3.1}$$

Responding, B, is controlled by its reinforcement, R, relative to total reinforcement from all sources, ΣR, whether associated with B or not. Interpretation of the terms in Equation 3.1 has varied, since it has been applied to data on choice, on simple response frequencies under positive and negative reinforcement, and on at least some of the phenomena grouped under contrast effects (Chapter 2; de Villiers, 1977; Herrnstein, 1970; Herrnstein and Loveland, 1974, 1975). The variables, B and R, typically refer to rates of behavior and reinforcement per unit time, as adapted for the particular circumstances being described. (See Chapter 2 for a variety of adaptations.) The parameter, k, and the summation, ΣR, are discussed in the next two sections. On the assumption that Equation 3.1 describes a body of data at least approximately, the purpose here is to derive it from simple and plausible preconditions. The plan of attack is to express underlying conceptions of behavior and reinforcement as factors in a differential equation whose solution is the matching law.

Invariance

It is postulated that reinforcement serves only to govern the distribution of behavior over the alternatives and nothing more, as regards its effects on frequencies of behavior. This means that when an increase in reinforcement produces an increase in the frequency of one alternative, other alternatives must decrease in frequency so as to hold constant the total amount of behavior. The obvious way to preserve invariance is to express behavior in units of time. Then the total amount of behavior equals the number of time units in a period of observation, however they were apportioned across alternatives. This is implicitly Premack's (1965; 1971) method and explicitly the method of Baum and Rachlin (1969) and Mazur (1977). Other measurement procedures may exist for preserving invariance in total behavior, although to my knowledge no one has formulated one. The invariance requirement is given by the equation:

$$\Sigma B = k. \tag{3.2}$$

The assumption of invariance, it should be emphasized, imposes a constraint on how behavior should be measured in order to show matching across differing topographies of response. Simple counts of varying response forms in asymmetric choices would usually violate the assumption: A rat traversing a 10-foot (3.05 m) runway is likely to execute fewer "responses" than one traversing a 5-foot (1.52 m) runway, for a given amount of reinforcement, if a complete transit from start box to goal is used as the measure of response. And indeed, untransformed counts of dissimilar responses (Chung, 1965a; McSweeney, 1978) do fail to equalize relative frequencies of response and reinforcement. The invariance assumption postulates that it is in principle possible to cut through differences in topography to a level of behavior measurement at which the effects of reinforcement leave total behavior unchanged, and that, at such a level, matching would hold. At present, this is offered as a presupposition, although measuring behavior in terms of response durations may satisfy the invariance requirement and there may or may not be other ways to meet it.

Different topographies of behavior may differ in two ways within the framework of the matching law. They may have different values of k associated with them, as illustrated in the runway example, but different topographies may also differ in the reinforcing value of the stimuli arising in their emission (Herrnstein, 1977a, 1977b; Herrnstein and Loveland, 1972). Traversing a 10-foot runway may well differ in inherent reinforcing value from traversing a 5-foot runway, simply because it requires more effort.

It has been possible to disregard differences in inherent reinforcement in most of the literature on matching because the topographies used in operant research appear to be close to zero on the scale of inherent reinforcing value under typical motivational conditions. Osborne (1977) summarized the evidence showing that pigeons and rats are essentially indifferent to whether or not they must peck a key or press a lever to earn food. We may, however, assume that they would not be indifferent if the response were something inherently reinforcing or aversive (i.e., negatively reinforcing). But such significant reinforcing values in a topography would be expressed in the value of R, not k, in Equation 3.1. The quantity expressing the obtained reinforcements for a response implicitly sums the extrinsic and intrinsic reinforcers contingent upon it, though the latter can be ignored with impunity in the typical experiment. Nevertheless, we should distinguish between the value of k associated with a topography and any intrinsic reinforcement accompanying its occurrence. The value of k should be interpreted strictly as a constraint on the measurement of behavior such that it remains invariant across changes in allocation among the different topographies, irrespective of the intrinsic positive or negative reinforcing values of the topographies.

When behavior is measured invariantly, the ratio between all behavior, k, and all reinforcement, ΣR, defines a quantity that figures in the derivation: $k/\Sigma R$. The factor quantifies the exchange relationship between behavior and reinforcement. Since total reinforcement is in principle unbounded and total behavior is assumed to be invariant, there must be more behavior per reinforcement the less the total reinforcement, and vice versa.

Relativity

The effect of a change in reinforcement on behavior is postulated to depend on how much of the subject's reinforcement is already associated with the behavior whose reinforcement is changing, just as if there were a Weber's law for reinforcement. When a large fraction of the total reinforcement is already associated with the behavior, an increment in reinforcement is assumed to have a smaller effect than if only a small fraction is already so associated. This conception is superficially similar to the economist's notion of diminishing marginal utility, but as will be shown, differs from it at a deeper level of analysis. In this essay relativity is expressed by the complement of relative reinforcement, that is, $1 - R/\Sigma R$. Note that this quantity varies between 0, when a behavior's reinforcement (R) is equal to all reinforcement (ΣR), and 1, when R is zero. It thus captures the Weberian idea of diminishing relative sensitivity (of a response to a change in reinforcement in this instance) as the response's relative level of reinforcement rises. In the derivation, we will consider the summation ΣR, to comprise R plus a constant background level of reinforcement, R_e. R, it should be recalled, is defined as reinforcement associated with the response in question, measured by B; R_e is defined as reinforcement not associated with B.

Invariance and Relativity Combined

Assuming continuity, the relation between a change in behavior (dB) and a change in reinforcement (dR) can be expressed in a differential equation combining the relativity and invariance factors. The form of combination used in Equation 3.3 is simple multiplication. Other forms of combination could have been employed, as long as they guarantee that the effect of an infinitesimal change in reinforcement is weighted by both the invariance and relativity factors, that the effect goes to zero when either factor goes to zero, and that behavior is monotonically related to each factor. Among acceptable combination rules, the simplest one is multiplication. The resulting differential equation is:

$$dB = \left(\frac{k}{\Sigma R}\right)\left(1 - \frac{R}{\Sigma R}\right) dR. \tag{3.3}$$

An infinitesimal change in reinforcement for a response is hereby postulated to cause an infinitesimal change in the response in accordance with Equation 3.3. From inspection, it is obvious that infinitesimal changes in reinforcement produce ever smaller changes in responding as the level of R rises, ultimately going to zero as R reaches ΣR. To predict rate of responding as a function of reinforcement, we cumulate the effects of the infinitesimal changes by taking the definite integral whose solution gives B, which is the total change in the rate of B when its reinforcement is changed from 0 to R :

$$B = \int_0^R \left(\frac{k}{\Sigma R}\right)\left(1 - \frac{R}{\Sigma R}\right) dR$$

$$= \int_0^R \left(\frac{k}{R + R_e}\right)\left(\frac{R_e}{R + R_e}\right) dR. \tag{3.4}$$

Bearing in mind that R is the variable and R_e is constant, the solution of Equation 3.4 is:

$$B = \left(\frac{-kR_e}{R + R_e} + c \right) - (-k + c)$$

$$= \frac{kR}{R + R_e}. \tag{3.5}$$

Equation 3.5 is the matching law as given in Equation 3.1. Matching is thus derived from a relativistic conception of reinforcement, given invariance in the measurement of behavior.

In one respect the foregoing derivation may seem seriously unrealistic, and that is the assumption of constancy in the background reinforcement R_e. In many settings, background reinforcement may in fact not remain constant as the response in question varies for any reason. For those settings, the matching law would not hold without correction. The survey of the literature in Chapter 2 on absolute response strength showed ample confirmation of Equation 3.5 with R_e held constant, but they did not discuss the possibility that the 5%–10% of unaccounted-for variance is partly a result of violations of this simplifying assumption.

Premack (1959, 1965, 1971) describes reinforcement as a relationship between contingent behaviors (or between behaviors and the accompanying stimuli) rather than as the absolute properties of stimuli. Premack and his associates showed that a reinforcing event in one context may be neutral or punishing in another and vice versa. This principle may seem to be relativistic in a different sense from the present one, which says that reinforcing values are scaled in terms of the total context of reinforcement (i.e., ΣR). Yet, both Mazur (1975) and Donahoe (1977) have demonstrated the formal similarity of Premackian relativity and the matching law. It is, however, just a similarity, not an identity. The empirical exceptions to Premackian relativity demonstrated by Mazur (1975, 1977) and others (Allison and Timberlake, 1974; Eisenberger, Karpman, and Trattner, 1967; Timberlake and Allison, 1974) may or may not be violations of the matching law depending on unresolved points of interpretation (Mazur, 1977). Moreover, Donahoe (1977) has suggested that Premackian relativity implies linear-operator models of acquisition, whereas according to the present analysis, the matching law is nonlinear with respect to acquisition, as will be shown in the next section.

A substantial body of data (reviewed in Chapter 2; de Villiers, 1977; Myers and Myers, 1977) supports the matching principle at least as an approximation for steady-state responding under the control of reinforcement. Given the logical bridge from Equation 3.3 to Equation 3.5, we can say that data showing matching (Equation 3.5) also show invariance and relativity in the effects of reinforcement (Equation 3.3), given that R_e, reinforcement not associated with the response in question, can be taken as constant. For independent confirmation of Equation 3.3, more data showing matching would be redundant. We must, instead, turn to other kinds of data from those that suggested Equation 3.5 to begin with.

Acquisition Curves

Matching is usually tested against the steady state of reinforced behavior. But if we assume that behavior during acquisition is also guided by reinforcement, then Equation 3.3 should be applicable to learning curves. Without taking a position on whether learning itself is dependent on reinforcement, we may assume that performance is, both in the steady state and in acquisition.

Mazur and Hastie (1978) surveyed the literature on simple, negatively accelerated acquisition curves for human subjects, such as those observed in free recall learning or simple perceptual-motor or manual dexterity learning. They present a case for concluding that most simple learning curves fit the hyperbolic form of Equation 3.5 better than they fit the more familiar exponential function of linear-operator models. Mazur and Hastie also point out that hyperbolic equations have been suggested in earlier works (e.g., Gulliksen, 1934; Thurstone, 1919, 1930). It may not be obvious how the measures of correct responding and practice in learning models can be equated with those of response rate and reinforcement in Equation 3.5, if indeed they can be. To do so presupposes that cumulative measures of practice are interchangeable with reinforcement for the correct response, that the learning rate parameter may be interpreted as reinforcement for alternatives to engaging in the correct response, and that the typical measures of correct responding in acquisition are analogous to those of response strength in the steady state. To the extent that Equation 3.5 describes acquisition, the foregoing presuppositions may seem more credible. It may therefore be useful to show in more detail how the present analysis and the linear operator model differ at the incremental or trial-by-trial level.

Linear models are usually expressed in terms of the probability of occurrence of a particular response on trial $n + 1$, that is, p_{n+1}, after reinforcement on trial n, as follows:

$$p_{n+1} = p_n + \Theta(1 - p_n), \tag{3.6}$$

where p_n is the probability of response on trial n and Θ is the learning-rate parameter. In the simplest models, Θ is a dimensionless constant, sometimes interpreted as measuring the proportion of a stimulus set sampled on any trial (Estes, 1959). Equation 3.6 is a linear model in the sense that it expresses the relation between behavior on two consecutive trials as a linear equation. It is easy to show that acquisition curves must follow an exponential form if individual trials obey Equation 3.6 (Restle and Greeno, 1970).

Equation 3.6 may be compared with the present analysis after several minor algebraic and notational adjustments. Equation 3.6, for instance, deals with response probability on trial $n + 1$, whereas Equation 3.3 deals with infinitesimal changes in responding, dB, arising from infinitesimal changes in reinforcement, dR. We can, however, solve for a finite change in responding, ΔB, associated with an increment in reinforcement from R to $R + \Delta R$. Applying Equation 3.5 to obtain the difference:

$$\Delta B = B_{(R+\Delta R)} - B_{(R)}$$

$$\Delta B = \frac{k}{R + R_e + \Delta R} \left(1 - \frac{R}{R + R_e}\right) \Delta R. \tag{3.7}$$

Here ΔR is taken as the reinforcement added to the cumulated total associated with practice on each trial and ΔB is the corresponding change in responding. It should be borne in mind that R is interpreted here as interchangeable with cumulated practice up to the point in acquisition at which ΔB is added.

Equations 3.6 and 3.7 are not yet suitable for direct comparison. In Equation 3.7, the term $R/(R + R_e)$ can be replaced by B/k, since Equation 3.5 holds. In Equation 3.6, we subtract p_n from both sides to get the change in probability of response, Δp, on any trial as follows:

$$\Delta p = \Theta(1 - p_n). \tag{3.8}$$

Finally, since k is by definition total responding, we note that the probability of response, p, is B/k and the change in probability of response on any trial, Δp, is $\Delta B / k$, in accordance with a commonly assumed relation between rate and response probability (see, e.g., Estes, 1959).

We may now rewrite Equations 3.7 and 3.8 to make them directly comparable:

Matching model (Equation 3.7):

$$\Delta B = \frac{\Delta R}{R + R_e + \Delta R}(k - B).$$ (3.9)

Linear-operator model (Equation 3.8):

$$\Delta B = \Theta(k - B).$$ (3.10)

Both models imply that increments in responding, ΔB, are inversely related to the amount yet to be learned $(k - B)$. For the linear model, it is an inverse proportionality; for the matching model, the amount yet to be learned is multiplied by $\Delta R / (R + R_e + \Delta R)$, a quantity that is itself inversely related to cumulated practice, R. Inasmuch as the multiplier interacts with the level of practice, the matching model resembles Mackintosh's (1975) theory of selective attention in acquisition, though Mackintosh has not applied his model to matching in the steady state and it is not clear whether the direction of interaction is the same in the two formulations.

Where R is small relative to R_e and where ΔR and R_e are themselves constant, the matching model becomes indistinguishable from the linear model. The expression $\Delta R / (R_e + \Delta R)$ would correspond reasonably well to Estes's interpretation of Θ as a proportion of stimulus elements conditioned to the correct response on any trial. However, with further practice, as B approaches k (i.e., p approaches 1), the two models must make divergent predictions because B cannot approach k unless R becomes large relative to R_e (see Equation 3.5). Mazur and Hastie (1978) found that the superiority of Equation 3.5 in comparison with the comparable exponential equation tends to appear late in practice. Compared to the matching equation, the best fitting exponential typically rises too rapidly with practice and then turns toward too low an asymptote for the data, a result, it would seem, of the difference between a fixed value of Θ and the decreasing value of $\Delta R / (R + R_e + \Delta R)$ with practice.

In summary, Equation 3.3 yields a model of performance during acquisition that has several notable features. It appears to describe learning curves well by usual standards. Although it is mathematically simple, it is nonlinear as opposed to linear-operator models. It thus differs not only from the linear models of Bush and Mosteller (1955) and Estes (1959), but also from their more contemporary descendants, such as Rescorla and Wagner's (1972) model of associational learning, Blough's (1975) model of stimulus discrimination and generalization, and Staddon's (1977) interpretation of matching itself. In addition, it does not have the property of path independence (Luce, 1959; Sternberg, 1963), because it includes a variable for cumulative practice. Whether the present analysis provides the useful or predictive stochastic properties of the earlier models cannot be said at present.

The Molecular Account of Matching

There has been some interest in deriving matching from more molecular processes, that is to say, from processes that control the emission of single, or just a few, responses rather than hundreds or thousands of them. The impetus comes in part from the hypothesis that the molecular processes may describe the subject's psychology better than the molar account. In extreme form, the hypothesis says that molar matching may be nothing but an averaging artifact, or as Shimp (1975) puts it: "Matching is an unimportant by-product of behavioral control by local processes" (p. 240). Silberberg et al. (1978) have recently expressed a similar view.

Shimp has said in a series of articles (e.g., Shimp, 1966, 1969, and summarized in 1975) that molar matching is the outcome of local (i.e., short-term) reinforcement maximization. Two arguments and related data are offered in support. First, matching is said to be the molar consequence only when local or momentary response strategies are optimizing local or momentary reinforcement probability. Second, it is said that when a procedure is specifically designed to yield discriminable changes in local reinforcement probability, the observed behavior patterns follow maximization strategies irrespective of molar matching.

The weight of evidence appears to support Shimp's second claim but not his first. Regarding the first claim, Shimp's data, as well as other data (e.g., Heyman, 1979; Nevin, 1969; Reynolds, 1963), contain instances of matching largely unaccounted for by sequential response strategies. Reynolds (1963), using both two- and three-key procedures with pigeons, found approximate matching by each subject. Reynolds searched without success for a rational, or even merely invariant, sequential pattern of responding. He concluded that despite the differences between birds at this level of analysis (i.e., sequential analysis) of their behavior, the invariance in these experiments is the overall equality between the relative frequency of pecks on one or two or three keys and the relative frequency of reinforcement obtained by pecks on that key (i.e., matching). (p. 59)

Later, Heyman (1979) examined concurrent variable interval–variable interval (conc VI–VI) for sequential sequences and concluded that matching is associated with virtually random patterns of changeovers between the alternatives. As far as behavior on conventional conc VI–VI is concerned, Heyman's data seem to be conclusive evidence against the momentary maximizing hypothesis.

Nevin (1969) studied pigeons pecking at two keys, but used discrete trials instead of the free-operant paradigm to simplify the analysis of sequential dependencies in relation to matching. Averaging over his three subjects, the relative frequency of pecking at one key was .75, whereas the relative frequency of reinforcement was .76; that is, matching held within 1%. However, Nevin found no sequential dependencies between responding and local probability of reinforcement. Choices on individual trials approximated the molar relative frequencies of responding, even though the transitional probabilities of reinforcement for individual choices were shifting by factors as large as 20 to 1 or more (see Nevin's, 1969, Table 2, p. 879). Comparable insensitivity to local shifts in reinforcement probability turned up in an unpublished experiment of my own (data presented in de Villiers, 1977), also using discrete trials and showing overall matching. Silberberg et al. (1978) interpret their data as being in conflict with these findings, but Nevin (1979) has now countered their interpretation, reaffirming his original analysis.

The first question, it should be recalled, is not whether sequential sequences of responding ever track transitional probabilities of reinforcement, but whether matching is ever observed in the absence of such response sequences. To that question, the answer is clear. An assortment of studies (e.g., de Villiers, 1977; Heyman, 1979; Nevin, 1969, 1979) has failed to support the claim that momentary or local maximization necessarily accompanies matching. In contrast, the claim that local contingencies of reinforcement sometimes produce response strategies that preclude molar matching appears to be on solid empirical ground. However, it should first be noted that matching predicts exclusive preference across choices that reinforce responses with different probabilities (Herrnstein, 1970; Herrnstein and Loveland, 1975), assuming that the subject has discriminated the difference. Therefore, when a subject discriminates shifting local reinforcement probabilities, as in some of the data reviewed by Shimp (1975), the matching law applied locally predicts local exclusive preferences, presumably for the alternative that is locally reinforcing with the higher probability. Molar matching may be violated, but the violation is trivial (see Herrnstein and Loveland, 1975), since matching is localized for each set of reinforcement probabilities. With his first claim refuted, Shimp's second claim may be correct, but it is no longer a molecular account of matching.

Relativity and the Law of Diminishing Marginal Utility

Earlier it was noted that the relativity factor in Equation 3.3 superficially resembles the law of diminishing marginal utility of neoclassical economic analysis in its implication that increments of reinforcement have decreasing effects as the level of reinforcement rises. According to price theory (Lipsey and Steiner, 1972; Stigler, 1966), the utility of successive units of a commodity must eventually decrease. The second banana may taste as good or even better than the first, but there must come an n beyond which successive bananas pall, says neoclassical economics.[1] From this law, in combination with other premises, economists derive the fundamentals of a marketplace—in particular, the determination of price, the allocations of income, and the maximization of utility.

Economists (e.g., Stigler, 1966) have justified the law of diminishing marginal utility in one of two ways. They may appeal to intuition as I just did in regard to bananas. The second way is a related argument based more directly on behavior: If marginal utilities did not inevitably diminish, then once a person ever preferred a commodity, the person would suffer the risk of gradually having all of his or her wealth tied up in increasing amounts of that commodity. For example, if a banana was what a person wanted at a given price and bananas did not pall at all, the person might decide to buy nothing but bananas. Since people rarely if ever behave so maladaptively, it is reasonable to assume that marginal utilities must eventually decrease.

Both of these economic justifications are based, from a psychological standpoint, on a diminution of drive, leading to a reduction of reinforcing power of stimuli (e.g., bananas); the first is based directly on satiation, the second, indirectly. In contrast, the relativity factor used here

---

1. We can overlook here the qualifications on this claim, which mainly concern the context in which the commodity is consumed.

has nothing to do with drive. It says that at a given moment, an increment of reinforcement is quantified by the organism with respect to the context of total reinforcement, irrespective of shifts in motivation.

The relativity factor embodies familiar ideas in psychology, even aside from the relativistic conceptions of reinforcement cited earlier. First, it says that the effect of a given change in R is inversely related to the initial size of R. Second, it says that the organism has no absolute scale of reinforcement, since the effect of a given change in R is proportional to units defined by the prevailing context of reinforcement, $R + R_e$ (see Equation 3.3). A similar conception applied to perceptual continua in general goes under the heading of adaptation-level theory (Corso, 1967; Helson, 1967). Context-dependent perception also enters into contemporary accounts of perceptual constancies and their physiological bases (e.g., Cornsweet, 1970). The present analysis may, therefore, be simply another example of contextual control over perception, in this instance the perceptual continuum of reinforcing value. Since the present analysis yields decreasing marginal reinforcing values independent of satiation or other changes in internal state, the economist's implicit reliance on changes in drive is superfluous for this purpose.

If the present substitute for diminishing marginal utility has nothing to do with drive, it is proper to ask how changes in drive fit into the theory. The economist's conception of shifting motivational states surely captures an aspect of the law of effect in nature. People and other organisms (see Pyke, Pulliam, and Charnov, 1977, for extensions of neoclassical economic analysis to foraging in animals) continually adjust current behavior to current utilities, which no doubt sometimes decline with consumption. Satiation and other motivational changes enter into the present analysis simply by changing the numerical values of reinforcers with respect to each other (Herrnstein, 1974a). As one overindulges in bananas, their value in an equation predicting behavior simply becomes smaller, passing into the region of negative values as banana approach gives way to banana avoidance.

Behavioral research has typically held drive states constant or almost so in studies of the law of effect. However, it is becoming clear that an important dimension of behavioral control emerges when the drives are allowed to fluctuate and seek a level as an organism interacts with its environment (e.g., Allison, 1976; Rachlin et al., 1976). To that extent, economic analysis provides a framework for psychological research. However, the present theory concerns the process governing an organism's behavior at each moment, given a certain internal state. As its internal state fluctuates, the process continues to govern behavior. In

contemporary applications of economic theory to the analysis of behavior, the momentary process is assumed to be momentary utility
maximization. The arguments against momentary maximization given
in relation to Shimp's account of matching apply equally well to these
applications. The present analysis suggests that one momentary process
governing behavior is described by Equation 3.3, the differential form of
the matching law.

Even though diminishing marginal utility in the economic sense is no
doubt often at work, nothing appears to require it as a universal element
in theories of behavior or choice. Female mammals in many species, for
example, place a high value on male companionship for relatively brief
periods of time. In cyclical species, such as the rat, most of the variance
in the "utility" of the male of the species to the female is probably accounted for by endogenous fluctuations of the female's cycle, not deprivation or satiation reflecting rates of consumption. Economists deal with
such complications by saying that they require constancy in "tastes" in
order to apply the neoclassical microeconomic analysis. Presumably, the
female rat's tastes would be said to be changing during her estrous cycle,
although the taste for bananas is assumed to remain unchanged as the
economist's law of diminishing marginal utility shrinks their marginal
value. But it is not clear why fluctuations in reinforcing power due to
cycles (or to shifts in light and temperature levels, etc.) are changes in
tastes, whereas the gradual cloying of bananas is not. The present analysis avoids arbitrary distinctions between tastes and other motivational
states. Drives, by definition (Herrnstein, 1977a, 1977b), govern the reinforcing powers of stimuli. At each moment, an organism's behavior
expresses the value of an equation that includes quantities for all ongoing reinforcement. As far as the law of effect is concerned, drives affect
behavior via those quantities and in no other way.

If choice does not always involve diminishing marginal utility in the
economist's sense, a question arises whether behavior may sometimes
reach equilibrium in the absence of overall maximization (interpreted as
total reinforcement). The answer is yes. Heyman and Luce (1979a) have
shown not only that nonmaximization is theoretically consistent with
matching, but that the widely observed matching on concurrent variable-
interval schedules may be an instance of nonmaximization under typical
conditions. Heyman and Luce show that reinforcement maximization on
conc VI–VI probably requires significant *over*matching to programmed
reinforcement rates, which has not in fact been observed (Heyman and
Luce, 1979b). The two attempts (Rachlin et al., 1976, Figure 7, p. 149;
Shimp, 1969, Figures 1 and 2, pp. 103 and 104) to simulate matching on

conc VI–VI by assuming that the subject is maximizing overall reinforcement both show deviations between matching and maximizing. Since the deviations were small, Rachlin et al. and Shimp disregarded them, but Heyman and Luce's derivation suggests that deviations (though often very small ones) between matching and overall maximizing in conc VI–VI are inherent in the procedure, not merely imperfections in the simulations. Herrnstein and Heyman (1979) have further shown that on conc VI–VR, the difference between matching and overall maximizing becomes too large to be disregarded. Yet they find matching by pigeons within about the same limits as on conc VI–VI. Given the similarities in the data, it seems unlikely that the matching shown by other species, including human subjects (Bradshaw, Szabadi, and Bevan, 1976), is any different. We may tentatively conclude that overall maximization, when it occurs, is derivative and that matching is more nearly fundamental, rather than the reverse. However, Equation 3.3 shows that behavioral change is directly and monotonically related to change in reinforcement (assuming $R_e \neq 0$). When an increase in reinforcement results from an increase in the associated response, then Equation 3.3 becomes a feedback system that will drive the rate of responding upward toward k. The directionality stated by this equation is itself a form of maximization, although a weak one, since it applies not to overall reinforcement, but to the reinforcement for each response in turn, given a constant background of reinforcement.

Summary

The present analysis does not stipulate different molar and molecular accounts of behavior. It implies that when behavior is in equilibrium with reinforcement, the allocation of behavior across the sources of reinforcement obeys the matching law. Given any discriminated change in reinforcement, either motivational or situational, behavior simply redistributes itself so as to reestablish matching. The derivation given above shows that this behavioral equilibrium will conform to matching if the effects of reinforcement are relativistic and total behavior in a given time interval is treated as a fixed quantity, as it would be if it were measured as time spent responding. Matching may be easier to identify in molar measures of behavior than in molecular measures because of the inverse relation between sample size and variability, but nothing in the data suggests that it is an emergent property of large samples of behavior or an averaging artifact. Matching is a simple quantitative statement of the law of effect, within the limits set by the available data.

4 | Melioration as Behavioral Dynamism

The idea of maximization, deeply rooted in common sense, holds behavior to be normatively rational and adaptive. Behavior is depicted as seeking an equilibrium that maximizes something—be it total subjective utility, hedonic value, reinforcement, energy intake, or reproductive fitness—within limitations of memory and discriminative acuity as well as the limitations imposed by the environment. Each mixture of behaviors and their outcomes is viewed as a unique bundle, and the organism is supposed to select the best bundle, on whatever is the relevant dimension. Equilibrium is reached with a distribution of activities that cannot be detectably improved upon by a redistribution of choices; that is, the obtained outcomes are maximized. This is the central conception of economic theory, around which the formal structure of economic analysis has developed. Not surprisingly, an intuitively compelling idea like maximization permeates behavioral sciences besides economics, such as sociology (Homans, 1974), behavioral biology (Pyke, Pulliam, and Charnov, 1977; Schoener, 1969, 1971; and Krebs and Davies, 1978), and psychology (Rachlin, 1980; Rachlin, Battalio, Kagel, and Green, 1981).

I will argue here that this plausible, all but universally accepted, and rigorous conception of behavior should be replaced by the matching law. The argument rests primarily on data that support matching over maximization in tests where the two theories make different predictions. Secondarily, the argument will try to show that the matching theory,

Originally published in M. L. Commons, R. J. Herrnstein, and H. Rachlin (eds.), *Quantitative Analyses of Behavior*, Vol. II: *Matching and Maximizing Accounts* (Cambridge, Mass.: Ballinger Publishing Co., 1982), pp. 433–458. Preparation of this chapter was supported in part by Grant DA-02350 from the National Institute on Drug Abuse to Harvard University.

once grasped, is at least as compelling intuitively and as amenable to formal elaboration and evolutionary interpretation as maximization.

Like maximization, matching theory should say something about both behavioral equilibrium and change. At equilibrium, matching requires the ratio of the frequencies of any two behaviors, B_1 and B_2, to match that of their obtained reinforcements, R_1 and R_2, as follows:

$$\frac{B_1}{B_2} = \frac{R_1}{R_2}. \tag{4.1}$$

This assumes that B_1 and B_2 are measured on a common scale and likewise for R_1 and R_2 (Chapter 3; Herrnstein, 1974a). The equation, for steady-state behavior and reinforcement, has been at least approximately confirmed in numerous experiments on a variety of species (Baum, 1979; de Villiers, 1977; Herrnstein, 1970).

Despite the confirming data, Equation 4.1 leaves an explanatory vacuum into which many hypotheses have been drawn (Catania, 1973; Donahoe, 1977; Gibbon, 1977; Killeen, 1981, 1982; McDowell and Kessel, 1979; Myerson and Miezin, 1980; Nevin, 1979; Staddon and Motheral, 1978). What is missing is a matching theory of behavioral change: How does an organism come to match its distribution of choices to the obtained reinforcements? Until we know the dynamic process that yields matching at equilibrium, it will seem reasonable to argue, as indeed some do argue, that matching is fortuitous (e.g., Shimp, 1969; Silberberg, Hamilton, Ziriax, and Casey, 1978), a by-product of a more basic law of behavior. Rachlin and other maximizers have suggested that the process of maximizing sometimes yields matching at equilibrium, but not necessarily. By definition, however, maximizers believe that behavior at equilibrium always maximizes, within the particular constraints.

Is there a process that guarantees matching at equilibrium, a dynamic process that does for matching theory what maximizing does for maximization theory? The answer proposed in Herrnstein and Vaughan (1980) was that behavior shifts toward higher local rates of reinforcement. Given a repertoire with only two alternatives, for example, if B_1 earns a higher rate of reinforcement than B_2, then time allocation shifts toward B_1 in a relatively continuous manner. If some distribution of time between B_1 and B_2 earns equal local rates of reinforcement, then equilibrium has been reached. If B_1 earns more reinforcement than B_2 at all time allocations, then at equilibrium B_1 displaces B_2 altogether. This process, called melioration, differs from the process of maximization in requiring the organism to respond only to the difference between local reinforcement rates from individual behaviors. Maximization, in contrast, requires the selection of the biggest aggregation of reinforcement across behaviors.

It will be shown that quite different behavioral equilibria can result from melioration and maximization. The two hypothetical processes nevertheless resemble each other in various respects: Both are hedonistic, utilitarian theories, inasmuch as behavior is held to be driven by its psychological consequences. Hence both can claim descent from the commonsense idea of reward and punishment and from the formal law of effect. Also, both theories involve adaptation in the sense of bettering a state of affairs, although melioration is a more limited adaptation than maximization, as will be illustrated later. Finally both require knowledge of environmental feedback functions and internal motivational dynamics before any actual behavior can be predicted. The difference between the two theories is, in short, circumscribed, though often large in predicted behavior.

In this chapter, reinforcement (or R_i) should be understood as subjective value, so that it can properly be said that maximization means obtaining the maximum aggregate reinforcement available. Maximizers may object to the term *reinforcement*, for they are much concerned with the way value interacts with consumption and also with nonlinear interactions between commodities. *Utility* is the preferred economic term to express subjective value, rather than *reinforcement*, which may seem too easily confused with the object itself. A food pellet should, in fact, not be equated with a certain magnitude of reinforcement, when its reinforcing value depends on past consumption and the mix of other commodities being taken. To avoid needless extra terminology, however, *reinforcement* will serve here as the controlling consequence of behavior in both theories. Whereas maximization centers on R_T, the total reinforcement from a distribution of choices, melioration centers on R_D, the difference between local reinforcement rates for any pair of behaviors.[1] Local reinforcement rate, in turn, is the reinforcement actually obtained from an alternative, R_i, divided by the time allocated to it, t_i. Equilibrium between t_1 and t_2 is achieved when, in the following equation, R_D is zero:

$$R_D = \frac{R_1}{t_1} - \frac{R_2}{t_2}. \tag{4.2}$$

---

1. Both R_D and R_T are actually reinforcements per total available time (session duration). In the equations that follow, session duration, or Σt_i, is set to 1.

When $R_D > 0$, time allocation shifts toward t_1; when $R_D < 0$, time allocation shifts toward t_2. When $R_D = 0$, then

$$\frac{R_1}{t_1} = \frac{R_2}{t_2}. \tag{4.3}$$

Equation 4.3 is the matching law for time allocation. Moreover, given, as assumed before, that responding to the two alternatives is measured on a common scale of behavior,[2] Equation 4.3 further implies the matching law as stated in Equation 4.1.

Melioration should account for behavioral change if it is to fill the vacuum left by the conventional matching law for behavior at equilibrium. Even though most of the data relevant to matching are for behavior at equilibrium, tests of the process of melioration can often be inferred from steady-state responding. As for a comparison between matching and maximization, melioration implies that behavior maximizes total reinforcement, R_T, under two and only two conditions, as follows:

(1) When $R_D > 0$ at all values of t_1 (see Equation 4.2) and R_T is at a maximum when t_2 is zero;

(2) When $R_D = 0$ at the maximum value of R_T.

Under the first condition one alternative earns more than the other at all allocations, and maximizing is achieved because melioration causes that alternative to become prepotent. Under the second condition melioration yields maximizing because the local reinforcement rates happen, by nature of the reinforcement feedback functions, to become equal at the maximizing time allocation across a pair of alternatives. Under any condition other than (1) or (2), behavior at equilibrium should deviate from maximizing R_T, if melioration correctly describes the dynamic process governing behavioral change. Under all conditions behavior at equilibrium should obey the matching law. Data from both maximizing conditions and several nonmaximizing conditions are presented in the next section, which is intended to illustrate melioration in relation to maximization.

---

2. It is assumed that frequency of responding is proportional to time spent responding. When two responses are measured on a common scale, they bear the same proportionality to time. Under these constraints, Equation 4.3, for time matching, implies Equation 4.1, for response matching.

A Sample of Findings

Concurrent Variable-Ratio, Variable-Ratio

The simplest procedure to represent from the standpoint of melioration happens also to satisfy condition (1) and therefore to imply maximization. Moreover, conc VR VR, or an approximation to it, occurs commonly in nature. The intuitive appeal of maximization may have something to do with its appearance in environments as familiar and simple as concurrent ratio schedules. The other side of the coin is the evolutionary selection of melioration, which may have something to do with its being a process that yields maximization in common environments. The evolutionary issue is considered again later.

The essential feature of conc VR VR is that the alternative behaviors are reinforced on variable-ratio schedules, which fix a given proportionality between behavior and reinforcement.

$$\frac{1}{(VR)_1} = p_1 = \frac{R_1}{B_1}. \tag{4.4}$$

The reciprocal of the VR_i value gives the probability of reinforcement p_i. The local rate of reinforcement on each of the concurrent VR schedules depends on the programmed probability of reinforcement and the local rates of responding, b_1 and b_2, which, evidence shows, are likely to be nearly equal in the range typically studied. Melioration is governed by the local reinforcement-rate difference:

$$R_D = \frac{p_1 b_1 t_1}{t_1} - \frac{p_2 b_2 t_2}{t_2}. \tag{4.5}$$

Equation 4.5 says that on conc VR VR, the time spent at each alternative, t_i, is immaterial. Assuming that $b_1 = b_2$, choice will shift toward the higher probability of reinforcement. If the probabilities are equal, all allocations are at equilibrium. Any difference in the local rates of responding simply shifts the point of equilibrium, $R_D = 0$, toward a particular unequal pair of reinforcement probabilities. All these relations are implied by setting Equation 4.5 equal to 0.

For maximization only the numerators from Equation 4.5 should be considered; these give the total amount of reinforcement for any distribution of time,

$$R_T = p_1 b_1 t_1 + p_2 b_2 t_2. \tag{4.6}$$

Recall that $t_2 = 1 - t_1$, which is to say that the alternatives are exhaustive. To find the equilibrium implied by maximization, we must see how R_T varies with changes in allocation. From Equation 4.6 and the definition of exhaustive alternatives, it follows that

$$\frac{dR_T}{dt_1} = p_1 b_1 - p_2 b_2. \tag{4.7}$$

Total reinforcement rises as the subject shifts toward the higher product of reinforcement probability and local rate of responding; or, if local rates of responding are equal, toward the higher probability of reinforcement.

The identity of Equations 4.5 and 4.7 tells us that for conc VR VR, melioration and maximization are indistinguishable. This is shown graphically in Figure 4.1 (from Herrnstein and Vaughan, 1980). The left ordinate shows R_D; the right one, R_T. Except for the lined-up zeroes, the two y axes are unrelated. The two functions plot Equations 4.5 and 4.6

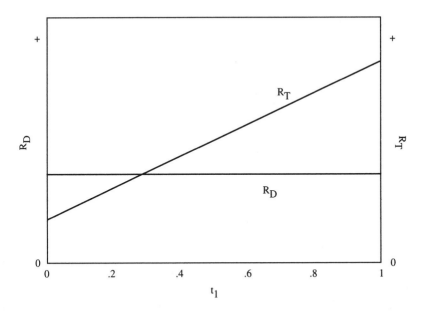

Figure 4.1   Concurrent VR VR in which alternative 1 reinforces with a higher probability than alternative 2. R_D is the local-reinforcement-rate difference between alternatives 1 and 2; R_T is the overall number of reinforcements. R_D and R_T are shown as functions of time spent responding on alternative 1, t_1, as a proportion of unit time. (From Herrnstein and Vaughan, 1980.)

against the respective ordinates. Figure 4.1 illustrates the case in which the probability of reinforcement for alternative 1 is the greater, and the local rates of responding neither change materially with time allocation nor differ from each other. Melioration predicts exclusive preference for alternative 1, since R_D remains above zero at all allocations. Maximization predicts likewise but for a different reason, which is that R_T, total reinforcement, is at its maximum at $t_1 = 1$. As both theories predict, subjects respond exclusively, or nearly so, to the better alternative when working on conc VR VR (Herrnstein and Loveland, 1975; Herrnstein and Vaughan, 1980) in which the component schedules are discriminably different. It should go without saying that exclusive preference on conc VR VR is no more evidence for maximization theory than for matching theory.

In general a comparison of matching and maximizing reduces to a comparison of the equation for R_D with that for dR_T/dt_1. Any environment that yields different expressions when the two equations are set equal to zero must result in a divergence between matching and maximizing. With conc VR VR the two expressions are identical (see Equations 4.5 and 4.7), hence the convergence of matching and maximizing.

Concurrent Variable-Interval, Variable-Ratio

In conc VI VR, one alternative reinforces on a variable-ratio schedule, with its preset proportionality between responding and reinforcement. The other alternative reinforces on a variable-interval schedule, which consists of a repeating series of irregular interreinforcement intervals. In Herrnstein and Heyman, 1979, a range of VR and VI values was explored to see where behavior reached equilibrium, particularly in relation to matching and maximizing. Since the results have been reviewed in several publications besides the original (e.g., Green, Kagel, and Battalio, 1982; Herrnstein and Vaughan, 1980; Heyman, 1982), the discussion here will be short.

Local rate of reinforcement behaves differently on the component schedules of conc VI VR. For VR it simply equals the product of reinforcement probability, p_2, and local rate of responding, b_2. For VI it depends (see Herrnstein and Heyman, 1979; Heyman and Luce, 1979a; for details and qualifications) on the scheduled average interreinforcement time V; the average interresponse time during responding on the VI, d_1; a measure of interchangeover times between the two schedules, I: and the proportion of time spent on the VI, represented as t_1 (proportion of time on VR is $(1 - t_1)$). Designating local rates of reinforcement for VI and VR as r_1 and r_2, respectively,

$$r_1 = \frac{1}{V + d_1} + \frac{1 - t_1}{V t_1 + I};$$ (4.8)

$$r_2 = p_2 b_2.$$ (4.9)

Note that r_1 depends on two terms, the second of which approaches zero as time spent on the VI approaches 100 percent. Thus, on VI there is an inverse relation between time spent at it and local rate of reinforcement. The formula for melioration on conc VI VR can now be stated.

$$R_D = \frac{1}{V + d_1} + \frac{1 - t_1}{V t_1 + I} - p_2 b_2.$$ (4.10)

Equilibrium is reached when R_D equals zero.

$$r_1 = r_2;$$

(4.11)

$$\frac{1}{V + d_1} + \frac{1 - t_1}{V t_1 + I} = p_2 b_2.$$

Since t_1 must fall within the $0 - 1$ interval, Equation 4.11 may or may not be solvable for t_1, given the other parameter values. If not, then melioration predicts exclusive preference for VI or VR, depending on whether R_D is always positive or negative, respectively. In the case of exclusive preference, matching is trivially satisfied. It can easily be proved that satisfying Equation 4.11 also implies matching, but not trivially. The number (as distinguished from local rate) of reinforcements from the VI, R_1, is the product of the local rate of reinforcement and the time spent there; likewise for VR, which is R_2.

$$R_1 = t_1 r_1$$ (4.12)

$$R_2 = (1 - t_1) r_2$$ (4.13)

However, at equilibrium (see Equation 4.11), $r_1 = r_2$. Hence, from Equations 4.12 and 4.13, time matching is the predicted (and observed) outcome.

$$\frac{t_1}{1 - t_1} = \frac{R_1}{R_2}.$$

Now let us consider maximization, which depends on the numbers (not local rates) of reinforcement:

$$R_T = R_1 + R_2. \tag{4.14}$$

From Equations 4.8, 4.9, 4.12, 4.13, and 4.14 it follows that

$$\frac{dR_T}{dt_1} = \frac{1}{V + d_1} + \frac{I - Vt_1^2 - 2It_1}{(Vt_1 + I)^2} - p_2 b_2. \tag{4.15}$$

Equilibrium would be reached when dR_T/dt_1 equals zero:

$$\frac{1}{V + d_1} + \frac{I - Vt_1^2 - 2It_1}{(Vt_1 + I)^2} = p_2 b_2. \tag{4.16}$$

If matching yielded maximization of reinforcement, then Equation 4.16 would be identical to the equilibrium produced by melioration (Equation 4.11), which it plainly is not. It can be shown that matching and maximizing can coincide only when responding to the VI stops, given realistic rates of alternation. Moreover, a comparison of Equations 4.11 and 4.16 indicates that matching generally requires more time spent on VI than on VR. Precisely that deviation toward spending time on the VI is the observed result (Herrnstein and Heyman, 1979).

The tendency to match on conc VI VR, as well as other features of the present analysis, is illustrated in Figure 4.2 (from Herrnstein and Heyman, 1979). Four pigeons worked on six pairs of values for conc VI VR. Steady-state responding is shown for 19 of the 24 conditions; the 5 points omitted had exclusive preference for the VI. The figure plots, in logarithms, the ratio of times on VI and VR against the reinforcements received from each, with the solid line tracing the best-fitting linear regression (of the logged variables) for the data pooled over subjects. Perfect matching is given by the dashed line. Except for a bias toward spending time on VI, which is itself consistent with the matching law (see Herrnstein and Heyman, 1979), the simple matching law accounts quite well for the data. In contrast, maximization, whose prediction is the dot-dash line, would have required substantial bias toward spending time on the VR. The optimal strategy for conc VI VR would seem to call for lots of time on the VR with occasional forays to the VI to collect a reinforcement come due, and analysis shows that intuition is right about the requirements of optimization for typical parameter values. Nevertheless, no subject displayed any such bias toward VR. Although it may be

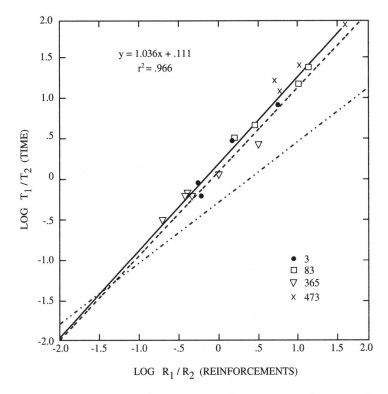

Figure 4.2   The logged ratio of times spent on the components of conc VI VR as a function of the logged ratio of reinforcements received from them, for four pigeons on six pairs of schedule values. The variable-interval component is designated by subscript 1. Simple matching is shown by the dashed line; maximization, by the dot-dash line. The best-fitting regression line through the data is the solid line; equation and percentage variance accounted for are also given. (From Herrnstein and Heyman, 1979.)

less obvious intuitively, optimal responding calls for *less* sensitivity to the distribution of obtained reinforcements than matching, shown by the lower slope of the maximizing prediction. The subject could earn more and think about it less, if it were a maximizer.

To show qualitatively how the two theories diverge here, a hypothetical conc VI VR is portrayed in Figure 4.3. The loci of Equation 4.10 for melioration and Equation 4.14 for maximizing are plotted against t_1, the proportion of time spent on the VI. For the values hypothesized, matching, where the local reinforcement rate difference R_D passes through zero, is for $t_1 = 0.45$; maximizing, for the same parameter values,

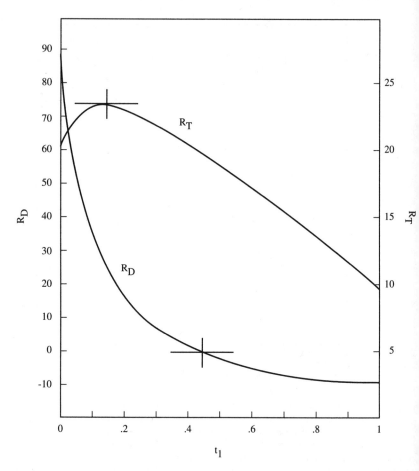

Figure 4.3 R_D and R_T for a hypothetical conc VI VR; t_1 gives the proportion of time allocated to the VI. Crosses show where R_D intersects zero and R_T passes through its maximum. (Figure and caption from Herrnstein and Vaughan, 1980.)

requires $t_1 = 0.15$. The value of R_T when $R_D = 0$ is about 15 percent lower than when R_T is maximized, which is the reinforcement cost of matching. In fact, by matching, the pigeons shown in Figure 4.2 gave up food reinforcements at a rate averaging about 60 per hour, a rate known to sustain nearly asymptotic rates of responding.

In spite of the apparently devastating outcome of conc VI VR, some maximizing theorists (Rachlin, Battalio, Kagel, and Green, 1981; Green, Kagel, and Battalio, 1982) have nevertheless held their ground. They point to the contribution of "leisure" as the reason subjects may sacrifice food reinforcement by overinvesting their time on VI. If a subject earns more total utility (or reinforcement, in our terms) by giving up food from the VR while replacing it with a lesser amount of food from the VI and a compensating gain from not working on VR (sparing it an aching beak?), then maximization may be saved. Just how this adds up has not been empirically demonstrated, although Green, Kagel, and Battalio (1982) believe they are on the track. But neither they nor anyone else as yet in print has even claimed to show why the leisure differential between VI and VR happens to be just the right magnitude so that subjects who appear to be matching on conc VI VR are actually maximizing. In any event, the ambiguity in conc VI VR arises from the difference in local response rates or in response topographies on the component schedules, which may indeed obscure an underlying maximizing process. The answer awaits the outcome of a more direct test of this possibility, which would be to devise a version of conc VI VR that eliminated or minimized the effect of topography or rate differences between the alternative responses. In the meantime, let us turn to two other recent experiments that confront matching and maximizing directly.

Mazur's Experiment

The procedure devised by Mazur (1981) (as well as that of Vaughan, described next) has the virtue of being so compelling intuitively that impenetrable formalizations are unnecessary. The pigeons could peck freely at either of two response keys, one green, the other red, in a standard chamber. A peck on one or the other key occasionally initiated 3 seconds of darkness. These periods of darkness were programmed by a single VI 45-second schedule, which randomly and equally often assigned each dark period to one key or the other, which makes this a Stubbs–Pliskoff (1969) procedure. Once primed, the VI timer stopped until a peck collected the assigned dark period. Hence the procedure guaranteed an exact, though random, sequence of switches in the effective peck from one alternative to the other. During a certain fraction of the 3-second dark periods, a small ration of food was delivered. For example, in the first condition every dark period was accompanied by food. In the second condition all dark periods for one alternative still yielded food, but only a random 10 percent of the dark periods for the other alternative did so. The six subsequent conditions presented further variations in the proportion of food-yielding dark periods.

Maximization and matching predict sharply different outcomes here. Consider the two conditions just described. For a maximizer the proper course of action was always to sample each alternative frequently and equally. Either key was equally likely to be primed to deliver a dark period, and the schedule only progressed as the assignments were collected, for both the first and second condition, as well as for all subsequent ones. For a matcher the tendency to equalize local reinforcement rates should have shifted preference along with the proportions of dark periods yielding food. In the first condition a matcher should have responded equally often to the two alternatives, just like a maximizer. In the second condition, however, when one alternative presented food 10 times as often as the other, a matcher should have spent over 90 percent of its time on the more lucrative alternative, even though doing so exacted a cost in food reinforcement. The cost arose because the VI timer stopped and waited at the end of each programmed interval as long as the subject was spending time on the unprimed side.

Mazur arranged four transitions for which a tendency to match would reduce overall food reinforcement (see Table 4.1 for the eight conditions), summarized in Figure 4.4. The foot of an arrow shows initial responding; the head, final responding during the decisive conditions. If pigeons were maximizers, they should have responded 50 percent to green under all conditions. Instead, in 16 out of 16 transitions (four for each pigeon), they shifted toward the matching value, which is indicated by the dashed vertical line. Fifteen of the 16 shifts caused reductions in food reinforcement, typically large ones. Unlike conc VI VR, here no problem of differential response rate or topography exists. Moreover, the

Table 4.1   Probability of food reinforcement in a dark-key period

	Key	
Condition	Red	Green
1	1.00	1.00
2	.10	1.00
3	1.00	1.00
4	1.00	.33
5	1.00	.00
6	1.00	.50
7	.50	1.00
8	.00	1.00

Source: Mazur, 1981.

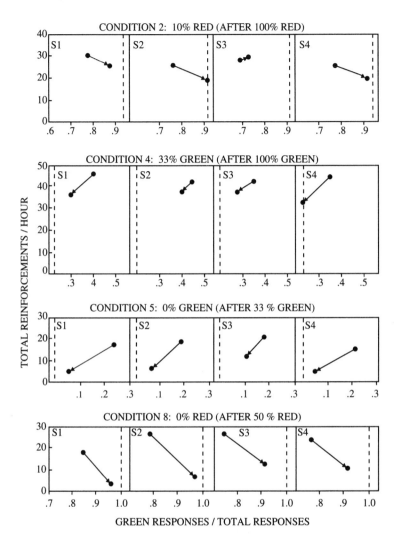

Figure 4.4 For four pigeons, the four transitions where matching predicts a shift away from equal responding to the green and red alternatives. All eight experimental conditions are summarized in Table 4.1. Overall reinforcement rate R_T is shown as a function of the relative frequency of responding to the green alternative. Nominal matching values are shown by the vertical dashed lines. The point at the tail of each arrow is the average of the first six sessions after a transition; the point at the head of the arrow is the average of the last six sessions before the next change of conditions. (From Mazur, 1981.)

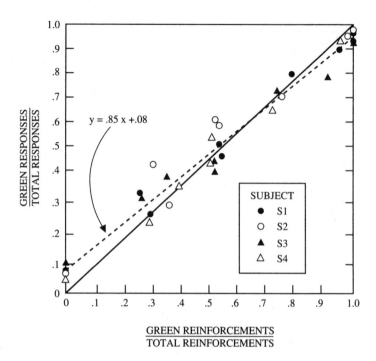

Figure 4.5   Relative responding to the green alternative as a function of relative reinforcement received therefrom, for each of four pigeons, averaging over the final six sessions of eight experimental conditions (see Table 4.1). Simple matching is the solid line; the best-fitting regression is the dashed line, whose equation is also given. (From Mazur, 1981.)

maximizing strategy could hardly have been simpler: whatever happens, respond half the time to each component. Figure 4.5, on coordinates of relative responding and relative reinforcement, shows the maximizing strategy as a dashed horizontal line, which the final performances of the four pigeons invariably failed to approximate. They approximated instead a function with moderate undermatching.

Mazur does not attempt to account for the undermatching, for his point was just the departure from maximizing, but two obvious reasons suggest themselves. First, the Stubbs–Pliskoff procedure itself may produce a characteristic type of undermatching because of the nonindependence of the two components. For example, even when the nominal reinforcement probability is zero for one alternative, responding to it is differentially reinforced (with some delay) because the responding advances the sequence of intervals upon which reinforcement from the other alternative depends. Second, the unusual contingency of reinforcement—a dark period that may or may not include reinforcement—may reduce the discriminability of the two alternatives, probably the most common source of undermatching. For these reasons Mazur's procedure can hardly be conducive to matching, although from the standpoint of simplicity it is ideal for maximizing. Nevertheless, the subjects approximately matched, shifting distributions of choices at considerable costs in reinforcement.

Vaughan's Experiment

Like Mazur's, Vaughan's (1981) procedure pitted matching against maximizing without any complication from differential response topography. To test the concept of melioration more directly than any previous study, Vaughan modified a concurrent variable-interval, variable-interval schedule in two respects. First the schedule values were continuously updated in reference to the subject's behavior, as will shortly be illustrated. Second, the two VI schedules advanced only when the subject was working on them, so that reinforcement for one alternative did not grow in probability with time spent on the other alternative, as in ordinary conc VI VI. When the subject was working on neither schedule, defined as any interresponse time greater than 2 seconds, neither schedule advanced.

The experiment consisted of two conditions, represented by a and b in the ensuing figures. Figure 4.6 shows the reinforcements per hour on the component VI schedules (L = left; R = right) as a function of relative time spent responding on the right. Conditions a and b are the same except when the relative time on the right falls between 0.25 and 0.75. In condition a, the left schedule then reinforces at a higher rate than the right schedule; in condition b, vice versa. Figure 4.6 shows how the subject's distribution of responding controlled the schedule values, which were updated every 4 minutes of responding. After every 4 minutes of responding the relative time on the right was automatically calculated. From that proportion the programming computer read off the appropriate values of the two component VI's to the program (as presented in Figure 4.6), which stayed in force at least until the end of the next 4-minute period of responding, when relative time on the right was again calculated.

Three pigeons ran on conditions a and b (except for minor additional adjustments in schedule values for one of them, described in Vaughan,

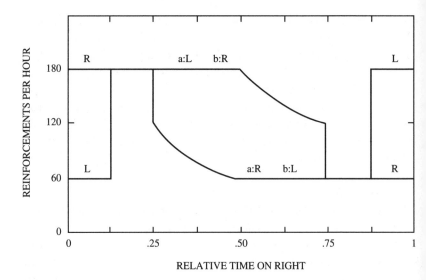

Figure 4.6 Programmed local rates of reinforcement on left (L) and right (R) alternatives as a function of time on right, measured as a proportion of unit time. The two experimental conditions, *a* and *b*, differ only for abscissa values between 0.25 and 0.75, as shown. (Adapted from Vaughan, 1981.)

1981). The point of the two conditions, hence the logic of the experiment, is highlighted in Figures 4.7 and 4.8. Figure 4.7 shows the maximization picture, namely how overall food reinforcements per hour interacted with the distribution of choices. In either condition, a subject earned the maximum, 180 reinforcements per hour, by spending 0.125–0.25 of its time responding to the right alternative. That is how a maximizer should spend its time. A sidelight that ought to be noted is that in both conditions, the minimum earnings, 60 reinforcements per hour, were given for spending 0.75–0.875 of the time on the right.

Figure 4.8 shows the melioration picture, namely the value of R_D at various distributions of choices. R_D, it should be recalled, is the local reinforcement rate difference between the two alternatives, here shown as "net reinforcements per hour on right," which is obtained by subtracting the left rate of reinforcement from the right:

$$R_D = \frac{R_R}{T_R} - \frac{R_L}{T_L}.$$

For condition a, the value of R_D stayed negative from 0.25 to 0.75 for time spent responding on the right; for condition b, it was always

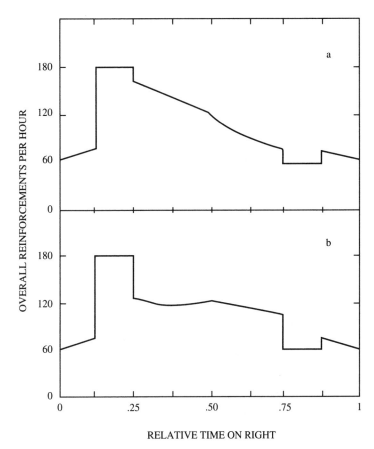

Figure 4.7 Overall reinforcements (R_T) as a function of relative time on the right alternative, for the two experimental conditions, *a* and *b*. (Adapted from Vaughan, 1981.)

positive in that range. Otherwise the two conditions were the same. Assuming that the central range was sampled, melioration should have held choice within the interval from 0.125–0.25 during condition a and within the interval 0.75–0.875 during condition b, without regard to the accompanying changes in overall reinforcement rate.

The predictions of the two theories and the experimental results are shown in Figure 4.9 for the three subjects. In condition a both theories predict choice in the range from 0.125–0.25 relative time on the right. Choice by the three subjects, shown by the points on the

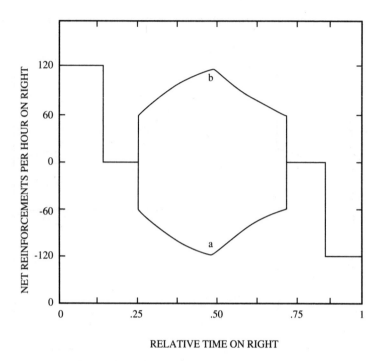

RELATIVE TIME ON RIGHT

Figure 4.8 Local rate of reinforcement on the right minus local rate of reinforcement on the left as a function of relative time on the right alternative. Conditions *a* and *b* differ only for abscissa values between 0.25 and 0.75. (Adapted from Vaughan, 1981.)

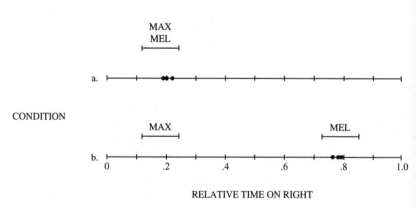

RELATIVE TIME ON RIGHT

Figure 4.9 Maximization ("MAX") and melioration ("MEL") predictions for conditions *a* and *b* in Vaughan's experiment (1981). The results for the three pigeons are shown on the scale for relative time spent responding on the right.

line, fell in the predicted range. After over 50 sessions in condition b, choice had stabilized in the interval 0.75–0.875, still meliorating but no longer maximizing. By meliorating, the subjects lost 120 reinforcements per hour, a drop of 75 percent in food intake. As in Mazur's procedure the dice were loaded for maximization, since no shift in responding at all from a to b would have earned maximum food. Vaughan (1981) noted, further, that in the shift from condition a to b, choice passed through distributions of behavior that transiently increased the deviations from matching, even though final performance in both conditions closely approximated matching, as it must if melioration is in control. This incidental finding rules out a dynamic process based on minimizing deviations from matching, a theoretical possibility entertained but rejected earlier in Herrnstein and Loveland (1975).

Limits of Melioration

Although the data make a clear case for melioration, as opposed to maximization, the theory's major limitations, empirical as well as theoretical, should be considered. The following discussion starts with the empirical questions, divided along the lines of standard psychological variables, and concludes with the inherent limits of melioration as a theory of behavioral adaptation.

Comparative

Most of the evidence for matching comes from pigeons, although there are enough data from other species, including human beings, to make matching a reasonably general principle of choice. However, the particular environments that most sharply separate maximizing from meliorating (Herrnstein and Heyman, 1979; Herrnstein and Vaughan, 1980; Mazur, 1981; Vaughan, 1981) have so far all been examined with pigeons as subjects. An argument could be made that since the equilibrium state—matching—has comparative generality, the dynamic process—melioration—must too, but this is an indirect argument. It remains conceivable that brighter creatures than pigeons do not meliorate but maximize, matching only when it involves no large cost in overall reinforcement. On the other hand, it is also conceivable that melioration can account for bright creatures as well as pigeons. Dealing seriously with comparative variation is a prime challenge to any general theory of adaptation and will therefore be reconsidered later.

Psychophysical

Melioration requires a subject to compare local rates of reinforcement, but this is "local" in a special sense. It is local with respect to response categories, not necessarily with respect to time. Pigeons, for example, compare left-key reinforcement rates to right-key reinforcement rates, but the time frame may be large or small. The rates are based, for each topography, on number of reinforcements divided by time invested in that topography, but we do not know all the parameters of the pigeon's estimation procedure. The transformation of objective local reinforcement rate into a subjective local reinforcement rate no doubt has its share of psychophysical complications (Commons, 1981; Commons, Woodford, and Ducheny, 1982). Species differences in this transformation may result in different equilibrium states even though each species is meliorating and matching.

Qualitatively, a subject's recent experience must weigh more than the distant past. Otherwise, a subject who spent 60 sessions on conc VI 1 minute VI 3 minutes would need more than 60 sessions to switch preference on conc VI 3 minutes VI 1 minute. Usually, preference switches rapidly, within 3 or 4 sessions, and the switch does not get materially slower with the duration of exposure to the preceding conditions, beyond some minimal exposure. From the standpoint of melioration this indicates a relatively rapid initial updating process for local rate of reinforcement, much as called for by Killeen's averaging theory (1981) or by the Rescorla-Wagner theory (1972) of associative strength (see Vaughan, 1982, for a concrete application to melioration). A subject's sensitivity to the current values of local rate of reinforcement should be reflected in the rapidity with which new equilibria are reached by melioration. One of the advantages of the melioration framework is its capacity to absorb a variety of equilibria.

The nature of the psychophysical transformation of objective rates probably explains a variety of findings that point to the importance of temporal structure above and beyond the average rates of events. It has, for example, been shown (Herrnstein, 1964) that pigeons prefer variable-interval to fixed-interval reinforcement when the objective rates are equal. Likewise, in their effects on behavior, remote reinforcers are less reinforcing (see Chapter 5; Commons, 1981; Gentry and Marr, 1980; Shull, Spear, and Bryson, 1981) and remote punishers less punishing (Hineline, 1977) than immediate ones, even after average reinforcement or punishment rate has been taken account of. Melioration theory presupposes nothing about the psychophysics of local reinforcement rate, only that behavioral equilibrium depends on what it happens to be.

Motivational

The four experiments described here were limited not only to pigeons as subjects but food as reinforcement. Choice between alternatives with the same reinforcer bypasses variations in demand elasticity, mutual substitutability, and diminishing marginal utility, complications well handled by microeconomic analysis. With only a few exceptions (e.g., Miller, 1976) matching theory has been applied to choices within single classes of reinforcers, from which some theorists (Hursh, 1978; Rachlin et al., 1981) have concluded that it cannot readily be applied across classes of reinforcers. However, as regards reinforcement interactions, there is no essential incompatibility with matching nor any essential compatibility with maximization. The difference is purely historical: Maximization theory has spelled out in great detail how motivational changes should affect choice, whereas it remains to be shown concretely how melioration and matching work in broader and biologically more relevant contexts.

Procedural

Matching was originally formulated as a description of choice, the relative frequency of a response with respect to concurrent behavior. Later it was extended to absolute frequencies of responding, which involved an interpretation of the context of reinforcement, and to behavior on multiple schedules, wherein the context of reinforcement shifts from time to time. In its present state of development, melioration has not been formulated or tested for anything but relative frequency of responding on concurrent schedules. The extension to single-response situations is in principle straightforward, however. For a single-response situation,[3] melioration presumably involves time allocation between the response in question, t_1, and all other activities t_e (such that $t_1 + t_e = t_T$, or total time) with their associated reinforcements, R_1 and R_e, respectively,

$$R_D = \frac{R_1}{t_1} - \frac{R_e}{t_e}.$$

---

3. The derivation assumes that the number of reinforcements R_i does not depend on time spent responding, as approximated, for example, by variable-interval or variable-time schedules. The assumption would also apply to discrete-trial paradigms in which some dimension of reinforcement other than frequency (such as amount or quality) was varied.

At equilibrium, $R_D = 0$, hence,

$$\frac{R_1}{t_1} = \frac{R_e}{t_e},$$

from which it follows that

$$t_1 = \frac{t_T R_1}{R_1 + R_e}. \tag{4.17}$$

Assuming that the frequency of the response in question, B_1, is proportional to the time spent at it, t_1, and that k measures total behavior in the units of B_1, it follows from Equation 4.17 that

$$B_1 = \frac{k R_1}{R_1 + R_e}. \tag{4.18}$$

Equation 4.18 is the familiar and widely supported hyperbolic equation (see Chapter 2) for absolute rate of responding as a function of relative reinforcement. Melioration can therefore be generalized from concurrent to single-response procedures. It is possible that a comparable argument can be developed for multiple schedules, but no fully satisfactory formula (including the one proposed by Herrnstein, 1970) for multiple-schedule responding has yet been demonstrated. It would therefore be premature to hypothesize further about melioration in multiple schedules here.

Inherent Limits

Melioration as so far described presupposes a set of response topographies with associated rates of reinforcement. But the acquisition of a new response topography itself exemplifies melioration, as less highly reinforced movements lose time allocation to more highly reinforced alternatives. The movements that survive exposure to a new contingency of reinforcement must, by this view, have equal local rates of reinforcement. But now a question arises. Is this class of equally reinforced movements also the class of maximally reinforced movements? If it is, then we must say that the end state of the process of melioration is maximization and the gap between the two theories would have vanished. If it is not, then why not?

As a concrete example, consider the conc VI VR shown in Figure 4.3, on which the shaping of response topographies need not have stopped with pecking at each of two keys. It could, by melioration, have continued until a more inclusive unit of responding was formed, one in which the subject spent 15 percent of its time on the VI alternative, which is the overall maximization point. In Mazur's and Vaughan's procedures, too, melioration could have led to maximizing compounds of the two response alternatives. The fact that this is not observed suggests that constraints on responding impede melioration in the formation of new topographies of responding. The constraints probably arise from several sources, such as the congeniality of certain movements as compared to others (which can itself be a component of local reinforcement rate), the organism's capacity to integrate reinforcement rates over longer or shorter time periods or to categorize certain movements as operants, and reversibility in the repertoire of responses.

For the perfectly plastic behavioral repertoire, melioration yields maximization. But to achieve perfect plasticity, the organism would have to be indifferent, motivationally speaking, to the physical characteristics of its own movements, capable of abstracting any contingency of reinforcement for any of its movements, no matter how subtle, unnatural, or over how large a time frame, and the effects of its past history of reinforcement would have to be totally reversible. Every distinct reinforcing environment would require a new configuration of movements as the maximizing topography, composed of simple movements that would lose their individual identity. To pick the right configuration the subject would also need something approaching perfect stimulus control. This ideal organism, which would maximize as it meliorated, may sound a bit like the hypothetical organism of early operant theory, with its "arbitrary" stimulus and response classes. Even if no such organism exists, as apparently it does not, it at least represents a benchmark for the adaptive potentialities of melioration. As evolution produces more easily conditionable and extinguishable response topographies and the capacity to detect correlations between behavior and its consequences over increasingly large time spans, melioration approaches maximization. Different species no doubt fall on different points along this continuum. Indeed, the ability to account for a continuum of adaptation that includes maximization as an extreme case makes melioration a likely dimension of evolutionary development.

Although maximizing theorists may sometimes neglect it, their theory, too, must have inherent limits. They do not believe that every behavioral configuration is perfectly geared into its consequences, now or later, near or far. Limitations of or defects in an organism's understanding, effects of uncertainty and time discounting are among the factors that spoil an organism's progress toward perfect maximization. In some respects maximization under realistic constraints may approximate melioration, with a drift toward higher local rates of reinforcement at the cost of more remote gains. Similarly the notion of momentary maximizing (Shimp, 1969) brackets melioration from the other side, by picturing the subject as controlled by the probability of reinforcement of the next response only. In spite of the convergence of these theories, a fundamental difference remains, however. According to melioration behavioral equilibrium must always conform to matching, no matter what the constraints of response topographies or cognitive limitations. The matching law cuts across the contingent limitations on melioration. For overall or momentary maximizing no universal characteristic of equilibrium has been specified except maximization itself, which is in fact an unobserved ideal case.

What Is Reinforcement?

A duality for reinforcement has long been customary. New topographies are held to be shaped out of the stream of behavior by the contingency of reinforcement, as in Thorndike's early puzzle box experiments (1898) and in countless other examples of *acquisition* via the law of effect. But then, after extensive study of intermittent reinforcement (Ferster and Skinner, 1957), it became clear that the *maintenance* of behavior is a salient second function of reinforcement. The present theory puts acquisition and maintenance into a single, quantitative conception. Melioration, the displacement of less highly reinforced topographies by more highly reinforced topographies, corresponds to acquisition. Given a set of topographies, melioration proceeds until behavior is in equilibrium, which is to say, at the levels dictated by the matching law, which corresponds to the maintenance function of reinforcement. Sometimes the equilibrium state also maximizes reinforcement and sometimes it does not, depending on the characteristics of the environment and of the organism interacting with it, as outlined in preceding sections of the chapter. The present theory is, then, a more precise and focused specification of the traditional law of effect and of common sense regarding reward and punishment.

If organisms are not literal but only accidental and occasional maximizers, one may ask why not. Why has natural selection followed the path of melioration, even though it leads to nonmaximization in certain environments? Given the evolutionary pressure toward more adaptive forms, why would there not be a maximization process in store for the future? The answers to these questions have been anticipated in earlier comments. Natural selection followed the path of melioration because it is a cognitively realistic approximation to maximization in many natural environments involving choices between probabilistic alternatives (that is, ratio schedules). Taking differences between local rates of reinforcement could well be within the capacities of organisms that could not aggregate reinforcement across topographies and over long periods of time, such as is required by maximization.

As to whether there is maximization in the future, an answer is that maximization is the limiting form of melioration. As an organism meliorates over varying response topographies, larger time frames, and a more differentiated set of environments, it approaches the ideal maximizer of economic analysis. Most economic theorists realize that their theory is built on a convenient fiction, that of maximization. The present analysis suggests that it is indeed a fiction for any existing organism, but hardly convenient if it obscures a more correct view of the matter. The law of effect interpreted as the dynamism of melioration leading to the equilibrium of the matching law appears to explain when organisms maximize and when they do not.

Part II

Self-Control

On its surface the matching law says nothing about the concept of self or the process of self-control. Yet it has profound implications for that concept and that process. The simplest sort of instance involving self-control poses choices such as the following: a pigeon choosing between one grain of food available one second from now and two grains of food available ten seconds from now; a child choosing between one piece of candy available one hour from now and two pieces of candy available ten hours from now; an adult choosing between a 13-inch television available one day from now and a 27-inch television available next month. In all of these cases one alternative is worth less but available sooner and the other is worth more but available later. Choice between them involves a tradeoff of amount (of the commodity) and delay; the higher amount goes with the longer delay and the lower amount goes with the shorter delay. People who choose the larger-later alternative are often said to be "controlling themselves" (in the sense that they forgo a smaller but still positively valued alternative). By the same token, if they choose the smaller-sooner alternative they are said to be "impulsive."

The original matching law says that measured degree of preference between two alternatives should be proportional to their relative rates of reinforcement. But matching has been generalized by Herrnstein and others to include all relevant parameters of reinforcement, not just rate. The ratio of the values of two activities is said to equal the product of the ratios of their reinforcement rates, their amounts, their immediacies (the inverse of their delays), the ease with which they are performed (the inverse of their efforts), and so on.[1] Where all parameters but amount and

---

1. The matching law says:

$$\frac{B_1}{B_1 + B_2} = \frac{R_1}{R_1 + R_2},$$

where B_i is a measure of behavior and R_i is the reinforcement rate dependent on that behavior. This equation is equivalent to:

immediacy are equal across alternatives, relative value is directly proportional to relative amount (A) and inversely proportional to relative delay (D). Thus the matching law predicts choice of whichever alternative has the higher A/D fraction (units equal across alternatives). In the case of the pigeon choosing between one grain of food in one second (A/D = 1/1 = 1) and two grains in ten seconds (A/D = 2/10 = .2), the matching law predicts preference for the smaller-sooner reinforcer. Indeed, with such a choice, pigeons would invariably opt for the smaller-sooner reinforcer—their choice would be "impulsive." But pigeons are not impulsive in this sense with every pair of alternatives. The matching law says that they would prefer two grains of food ten seconds from now (A/D = .2) to one grain of food six seconds from now (A/D = .17)—and so they do.

But which alternative is *really* better? There is no fixed answer to this question; it depends on need. For example, only with an astronomical interest rate would the money saved by forgoing a 13-inch television grow enough in one month in your bank account to enable you to buy a 27-inch television. Given actual interest rates, a 27-inch set next month is worth much more than a 13-inch set tomorrow. Yet if you were given a choice between those two alternatives it would not necessarily be irrational to choose the 13-inch set tomorrow. (The World Series might be on next week.)

For an economist, rationality lies not in which alternative you choose but in whether you choose that alternative *consistently* over time. The pigeon that prefers two grains of food ten seconds from now to one grain six seconds from now should maintain its preference for the larger al-

---

$$\frac{B_1}{B_2} = \frac{R_1}{R_2}.$$

Other factors such as reinforcer amount (A_i), immediacy ($1/D_i$), or unknown factors such as the tilt of the Skinner box or attractiveness of the color of the keys (X_i) are assumed to combine multiplicatively:

$$\frac{B_1}{B_2} = \frac{X_1}{X_2} \cdot \frac{A_1}{A_2} \cdot \frac{D_2}{D_1} \cdot \frac{R_1}{R_2}.$$

If all factors except amount and delay were kept equal, then

$$\frac{B_1}{B_2} = k\frac{A_1}{A_2} \cdot \frac{D_2}{D_1},$$

where $k = X_1/X_2$.

ternative after a second has passed (two grains in nine seconds versus one grain in five seconds), after two seconds have passed (two grains in eight seconds versus one grain in four seconds), and so forth. Yet the matching law predicts that as time passes *the pigeon will change its preference* ("change its mind") and will eventually come to prefer the smaller-sooner alternative. After five seconds have passed, for instance, the two grains of food (originally ten seconds in the future) will have doubled in value from .2 (A/D = 2/10) to .4 (A/D = 2/5) but the one grain of food (originally six seconds in the future) will have grown nine times in value from .11 (A/D = 1/6) to 1.0 (A/D = 1/1). At this point, the pigeon is predicted by the matching law to be impulsive. Indeed in everyday human life such inconsistencies of choice are all too common. To quote Rex Stout (from *The Rodeo Murders*), "The trouble with an alarm clock is that what seems sensible when you set it seems absurd when it goes off." Exactly because it predicts such changes in prefer-ence, the matching law differs from any theory of behavior (such as those common in economics) which says that behavior (of humans or nonhu-mans) must always be rational. (The chapters in Part II make this point repeatedly.)

A classical economist would not deem you to be irrational for valu-ing one delayed reward less than another (much of economics is in fact based on temporal discounting), but an economist would say that you are irrational or at least inconsistent if, without new information, you change your mind at any point. Therefore the only (economically) con-sistent account of the effect of delay on choice is by a discount function that predicts no reversals of preference with time. The only function that has this (nonreversing) property is an exponential function. The match-ing law, which does predict reversals, posits value as a hyperbolic (an inverse), not an exponential, function of delay. Behavior is often incon-sistent in just this way, says the matching law.

In Chapter 5, with Shin-Ho Chung, Herrnstein shows how matching, previously applied to the overall rate of reinforcement, may be applied to an individual delayed reinforcer—the rate of which is the simple inverse of its delay. In a choice between two delayed reinforcers, the relative value of each is predicted by matching (all else being equal) to be in-versely proportional to the relative delays of each. The data reported by Chung and Herrnstein have been reinterpreted in various ways and the extent to which they actually support the matching law is in question. But we include the essay here because it was the first to derive temporal discounting from matching.

In Chapter 6, published fourteen years later, Herrnstein elaborates the matching law to account for two readily apparent deficiencies in its simpler (amount/delay) formulation. The first is that the value of an alternative given by the fraction, amount/delay, approaches infinity as delay approaches zero. This formulation predicts that any reinforcer amount, no matter how small, available immediately (delay = 0) is worth more than any amount, no matter how large, available with a finite delay. For example, a dollar immediately (a dollar that you have) would be worth more than a million dollars delayed by one second. Because such a preference would never be found, the simple A/D formula for value cannot apply to very small delays. An answer to this is that no choice alternative, no matter how quickly gotten, is ever gotten immediately (with zero delay). There is always some threshold of delay between acquisition and consumption of a reward. But this answer is unsatisfactory because it is extrinsic to the matching law. In modifying simple A/D matching Herrnstein's tactic was to divide the matching relation (originally between behavior and reinforcement) into two parts: first, reinforcement is converted to its subjective value; second, behavior is said to depend on that value. The modification Herrnstein presents in this essay resides in the reinforcer-value relationship. In the modified formula, when delay goes to zero, value does not go to infinity but is based on relative rate with no delay (reinforcement due to responding relative to all reinforcement available in the situation).

A second deficiency in the value = A/D formulation is that once amounts are converted to utilities (perhaps by the law of diminishing marginal value) and delays are converted to subjective delays (perhaps by Stevens's psychophysical power law), there is no further accounting for individual differences in impulsiveness. Yet we know that organisms with similar psychophysical functions of amount and delay differ vastly in impulsiveness. Children, for instance, are notoriously more impulsive than their elders. Much of modern developmental and personality psychology relies on differences in ability to delay gratification—with age, across cultures, within families, and so forth. Herrnstein's modification of the matching law in Chapter 6 adds a parameter (I) to incorporate such differences. Every individual could be assigned an I-value reflecting his or her overall tendency to be impulsive. As your I-value increases you tend (according to this amended version of the matching law) to choose larger-later alternatives more frequently.

Between the publication of Chapter 6 (1981) and Chapter 7 (1988), James E. Mazur performed an important empirical study of choice by pigeons among various amounts and delays of reinforcement. Mazur found the value of a delayed reinforcer to be given not by A/D but by A/(1 + KD) where K is a constant determining degree of sensitivity to delay. In Mazur's formulation, when delay is zero, value is simply equal to amount of reinforcement.

Chapter 7, by Mazur and Herrnstein, incorporates Mazur's formulation of value with matching in a more empirically based approach (than that of Chapter 6) remedying the two deficiencies of simple (A/D) matching already noted. First, in Mazur's formula, when delay is zero, value is simply a function of amount and does not go to infinity. Second, the constant K provides an empirical measure of sensitivity to delay that was found to vary between individuals while remaining constant for an individual as delay and amount varied. In the years since Chapter 7 was first published more and more evidence has accumulated in support of Mazur's formula.

Chapter 8 is a review of a collection of diverse essays on the concept of multiple selves. Here Herrnstein makes explicit his ideas on rationality and self-control. The Cartesian concept of self-control postulates an internal battle between reason and passion. Most of the reviewed articles discuss the concept of self in terms of this battle. A person's true self is traditionally said to be his rational self, and self-control is traditionally viewed as dominance of reason over passion. But behavior predicted by the matching law corresponds to rational behavior only in certain circumstances. Therefore Herrnstein cannot accept a view of the self based on reason. In this chapter Herrnstein says essentially that the concept of the self would be meaningful in psychology only if "self" were defined as "that which obeys the matching law and other fundamental behavioral laws."

Chapter 9, by Herrnstein and Dražen Prelec, is an ambitious attempt to apply matching principles to various self-control problems. In this article the matching law is used to develop what the authors call the *primrose path* theory of addiction. (The reason for the change in terminology from "self-control" to "addiction" is to directly contrast the matching law with economic theories of "rational addiction.") Herrnstein and Prelec point out that crucial self-control situations in everyday human life are not merely choices between individual sooner and later rewards. (In the very first essay included here Herrnstein stated that his goal was to investigate output as a function of *frequency* of reinforcement.) An alcoholic, for example, is not seen as just choosing to drink this drink or not to drink it; the alcoholic is seen as choosing a certain *frequency* of drinking—and in doing so, according to the (relativistic) matching law, choosing a complementary frequency of nondrinking. According to Herrnstein and Prelec, we calculate separately the value of each consummatory activity (divided by time spent at that activity) and choose whichever is higher. An alcoholic, for instance, might often find the value of drinking, thus calculated, to be higher than the value of not drinking. In general, however, due to habituation or satiation, the more time we spend at a given activity, the lower its overall value. Thus the alcoholic's repeated choices of drinking might lower its value below that of not drinking and the process may come to equilibrium. But for harmful addictions the path to equilibrium is a "primrose path" because not only does increased consumption of addictive substances reduce their consummatory value, but it also reduces the values of alternative activities. For the alcoholic, drinking is more valuable than not drinking at all but the highest possible drinking rates. But at the point of equilibrium the value of both alternatives, drinking and nondrinking alike, is very low (due to hangovers, incapacities, bad health, social antagonisms, and so forth). Our tendency to match keeps us focused on *relative* values. But by matching we lose sight of *absolute* values as all alternatives deteriorate together.

How then can the "primrose path" be avoided? According to Prelec and Herrnstein, the first step in such avoidance is a step back, by which we perceive the path without being on it, so to speak. We see the consequences of the primrose path in the behavior of other people—friends, family, newspaper accounts, biographies, fiction. Second, we expose ourselves to the perception of others, of society as a whole, and to the commitments and constraints of custom and law. For Prelec and Herrnstein the imposition of such commitments and constraints is one of the primary functions, if not the main function, of society.

5

Choice and Delay of Reinforcement

SHIN-HO CHUNG AND RICHARD J. HERRNSTEIN

The present experiment extends the investigation of reinforcement delay described by Chung (1965b). In the earlier study, pigeons were trained to peck two response keys and received food reinforcement on equated variable-interval schedules. The reinforcements for responses on one key were delayed by various durations of time out, while the reinforcements for responses on the other were immediate. The findings suggested a negative exponential relation between the relative frequency of responding and the duration of the delay. The present experiment also used a concurrent procedure, but differed in setting delays of various durations for both of the response alternatives.

Method

Subjects

Six male White Carneaux pigeons, experienced in a wide variety of experimental procedures, were maintained at approximately 80% of free-feeding weight.

Apparatus

A pair of experimental chambers for pigeons was used. Each chamber contained two response keys, spaced 9 cm apart, and a feeder providing 3-sec access to food for reinforced responses. Effective pecks had to be

Originally published in *Journal of the Experimental Analysis of Behavior*, 1967, *10*, 67–74. Copyright 1967 by the Society for the Experimental Analysis of Behavior, Inc. This work was supported by grants from the National Science Foundation to Harvard University and from the National Institutes of Health (Grant MH-12108-01) to the Foundation for Research on the Nervous System.

of at least 10-g force and each operated a relay to provide auditory feedback to the pigeon. The chamber was illuminated by a white bulb, and, except when the magazine was operated, each response key was transilluminated by a 7-w red bulb. A continuous white masking noise was delivered during sessions.

Procedure

Pecks on either of the two response keys were, at first, reinforced on a variable-interval schedule with an average interval of 1 min. Two independent programmers arranged reinforcements for responses on the two response keys, with the restriction that a switch from one response key to the other prevented reinforcement for 1 sec (changeover delay or COD 1-sec). When the rate of pecking on the two keys became stable and approximately equal, delays of reinforcement were initiated. For subjects 237, 236, 415, and 416, responses on the left key were reinforced after an 8-sec delay (standard key). Reinforcements for responses on the right key were delayed for various durations ranging from 1 to 30 sec (experimental key). The intervals of delay imposed on the experimental key, in irregular order, were: 1, 2, 4, 6, 8, 12, 16, 20, 24, and 30 sec. For the other two subjects, 242 and 211, reinforcements for responses on the standard key were delayed for 16 sec. The intervals explored on the experimental key were: 2, 4, 6, 12, 20, 24, and 30 sec. From 21 to 40 sessions were given for each pair of delay intervals, depending on how quickly stable performance was attained. Sessions were terminated after the 60th reinforcement.

Between the response-to-be-reinforced and delivery of the reinforcement, the chamber was darkened and responses produced no auditory feedback. As expected, this delay period contained virtually no responses. In Chung's previous study (1965b), reinforcement-delay periods were produced by responses on one key only, while responses on the other produced an equal number of time-out periods uncorrelated with reinforcement. This feature of his procedure, which was an effort to cancel out the effects of time out per se, as distinguished from the effects of reinforcement delay, was not duplicated here. One subject (S-326), not listed above, was disqualified from the experiment after initial training, since it responded only on one of two keys when any pair of delay intervals was imposed. This may have been due to this subject's prior exposure to an experiment in which reinforcements were followed by blackouts for one of the two response keys and no reinforcements were followed by blackouts for the other.

Results

The relative frequency of responding on the experimental key was found to be a joint function of the delay intervals on that key and on the standard key. With an 8-sec delay on the standard key, the relative frequency of responding on the experimental key varied from 0.82 to 0.15 as the delay interval varied from 1 to 30 sec. With a 16-sec delay on the standard key, the relative frequency of responding on the experimental key varied from 0.91 to 0.34 as the delay interval varied from 2 to 30 sec. In Figure 5.1, the relative frequency of responding on the experimental key is plotted against the duration of the delay. Points for each subject in Figure 5.1, and in the subsequent figures, were obtained by averaging the performances of the final 10 sessions at each duration. The upper curves in Figure 5.1 were obtained from the group receiving the 8-sec delay for the standard key; the lower curves are from the group with the 16-sec delay for the standard key.

The gross features of the two sets of curves are in agreement. In both instances, the general trend appears to be a monotonically decreasing function. However, there are certain conspicuous differences between the two sets of curves. First, at each value of delay, the height on the ordinate for the upper curves is, in almost every instance, less than the height for the lower. In other words, a smaller fraction of responses occurs for a given delay if the alternative is 8 sec than if it is 16 sec. Secondly, the evidence for upward concavity is more pronounced for the upper curves than for the lower, where the evidence is questionable at best. And, finally, the upper curves appear to pass through the value of 0.5 on the ordinate at close to the expected value on the abscissa—8 sec—whereas the lower curves are clearly elevated about 0.5 at 16 sec. Assuming that the pigeon would distribute its responses equally between the two keys when delay durations for the two keys were equal, the elevation of the lower two curves suggests that some degree of artifactual key-preference was present for the 16-sec group.

These various features of the individual functions are shown for the averaged groups in Figure 5.2, which also includes the data from Chung's 1965b study in which the pigeons (a group of three) were choosing between the delays shown on the abscissa and immediate reinforcement. The addition of this third function further substantiates the trends already noted. With presumably zero delay on the standard key, the degree of upward concavity is further accentuated. Moreover, this additional function is situated even lower on the ordinate than the other two. One further aspect of this added function might be noted. When

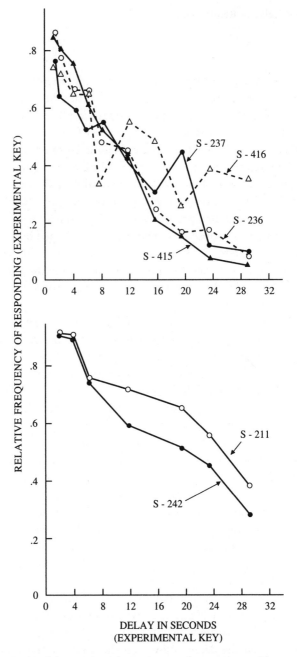

Figure 5.1 Ratio of the number of responses on the experimental key over the total number of responses on both keys as a function of the duration of the reinforcement delay for the experimental key. Each point is the average of 10 sessions for one subject. The upper curves are for the subjects with an 8-sec standard delay; the lower curves for the subjects with a 16-sec standard delay.

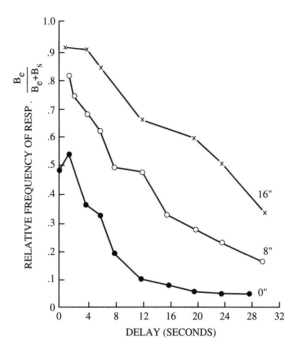

Figure 5.2   Relative frequency of responding on the experimental key as a function of the reinforcement delay for that key, averaged across subjects. The parameters refer to the duration of the standard delay. The bottom curve was taken from Chung (1965b).

responses on either key were reinforced immediately, *i.e.*, at zero on the abscissa, the pigeons responded equally, as would be expected. Contrary to expectation, however, the curve rises with a delay of 1 sec before it starts to decline along the exponential curve used by Chung to describe these data. This rise is probably genuine, for it was observed for each of the three subjects. Nor is the rise actually contrary to intuition, considering the actual circumstance. A 1-sec delay of reinforcement gives the pigeon time to get its head into, or close to, the feeder opening so that it may start eating as soon as the feeder arm is within reach. The putative "immediate" reinforcement, in fact, involves whatever delay is accounted for by the pigeon lowering its head to the feeder and probably involves a shorter effective reinforcement duration than a reinforcement delayed for 1 sec. It may, therefore, be entirely proper to consider the second crossing of the 0.5 level, at about 1.5 sec of delay, as the point at which the delays for the two keys were actually equal.

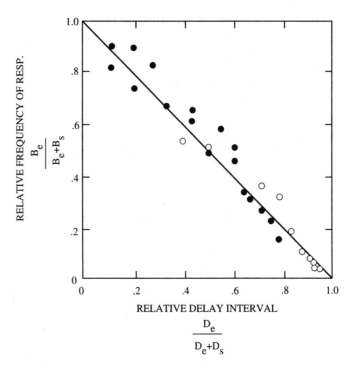

Figure 5.3   Relative frequency of responding on the experimental key as a function of the ratio of the duration of the delay for that key over the sum of the durations of the delays for the two keys. The filled circles are values for the 8-sec and 16-sec groups separately; the open circles are from Chung (1965b).

The relative frequency of responding at the experimental key ($\frac{B_e}{B_e + B_s}$) as a function of the relative duration of the delay intervals on the standard and experimental keys (i.e., $\frac{D_e}{D_e + D_s}$, with D_e for the experimental delays and D_s for the standard delay) is shown in Figure 5.3. The data obtained from the earlier study (Chung, 1965b) were included in Figure 5.3 by taking a small constant as the actual delay interval for what is nominally immediate reinforcement. The value of the constant (1.6 sec) was estimated so as to minimize the sum of the squared deviations between the observed values and the function predicted from the present findings. These earlier data are shown as open circles, presenting the average of three pigeons. The filled circles show separately the 8-sec and the 16-sec groups in the present experiment. Figure 5.3 shows that the relative frequency of responding closely matches the relative immediacy of the delay intervals, if relative immediacy is taken as the complement of relative delay. The more familiar increasing diagonal could just as well have been obtained using immediacy, defined as the reciprocal of delay, as the independent variable. Thus,

$$\frac{B_e}{B_e + B_s} = \frac{D_s}{D_s + D_e} \quad \text{or} \tag{5.1}$$

$$\frac{B_e}{B_e + B_s} = \frac{i_e}{i_e + i_s}, \tag{5.2}$$

where i is the reciprocal of D, and the subscripts distinguish the standard key from the experimental key. Although the data approximate the diagonal reasonably well, the 16-sec group tended to fall consistently above the diagonal, as would readily be predicted from the average curve in Figure 5.2. This deviation seems to be attributable to a key preference since only the intercept, and not the slope, of the function is affected.

Although the programmed rate of reinforcement for responses on the two keys was identical, the relative frequency of reinforcement actually delivered for responses on each varied systematically as a function of the relative delay interval because rate of responding decreased with increases in the delay interval. Figure 5.4 shows the relative frequency of responding as a function of relative frequency of reinforcement actually delivered for responses on that key. Once again, the open circles show the data from Chung (1965b). The function deviates systematically from the linear relation, indicating that the changes in the relative frequency of reinforcement do not sufficiently account for changes in the relative frequency of responding, and that delay itself is instrumental for the function in Figure 5.3.

The absolute rates of responding, as well as the reciprocal relation between the responding on the two keys, are shown in the three sections of Figure 5.5, corresponding to the three values of the standard delay studied in Chung (1965b) and the present experiment taken together. The filled circles show the average rate of pecking at the key with the varying duration of delay; the open circles, at the key with the fixed duration as indicated for each section. The curves were fitted by eye. Various features of these curves are readily noted. It is clear that a change in the delay value for one key affects the rate of responding on both keys; that the effect is opposite and, to some fair degree of approximation, symmetrical; that the amount of curvature in these functions decreases with an increasing standard delay, and that the point of intersection for each pair of curves is further to the right as the standard delay is

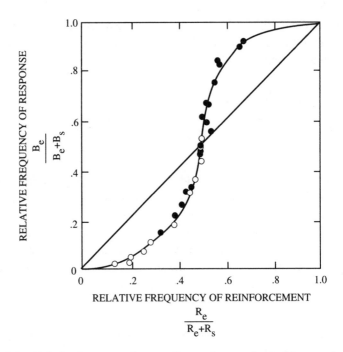

Figure 5.4   Relative frequency of responding on the experimental key as a function of the relative frequency of reinforcement associated with that key. Filled circles are for the 8-sec and 16-sec groups separately, open circles are from Chung (1965b).

increased. All of these features, as will be shown, are characteristic of behavior that obeys the matching relation depicted in Figure 5.3.

Discussion

The central fact disclosed by the present study is that the relative frequency of responding matches the relative immediacy of reinforcement in a two-response situation. Hence, delay of reinforcement may now be added to those other variables, like frequency and amount of reinforcement, whose effects are adequately depicted as the subject's matching of its behavior to the ratio of the magnitudes of the independent variable. Thus, the ratio of choices in a two-choice situation has been shown to equal the ratio of rates of reinforcement (see Chapter 1); to equal the ratio of amounts of reinforcement (Catania, 1963a); and, here, to equal the ratio of immediacies of reinforcement. But since Catania's experiment varied amount of reinforcement by varying the duration of feeder operation in a conventional key-pecking apparatus, it could be argued

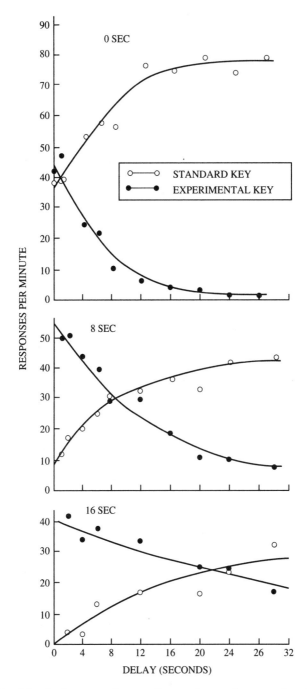

Figure 5.5 The absolute rate of responding on the two keys as a function of the delay for the experimental key, averaged across subjects. The three sections are for the three durations of the standard delay. Filled circles show rates for the experimental key; open circles, for the standard key.

that these three findings arise from the properties of the pigeon's perception of time and nothing more. Baum (1966), however, demonstrated essentially the same relativity of choice using rats, instead of pigeons, and sucrose concentration—in no obvious way a temporal parameter—as the independent variable.

The traditional interest in delay has not, however, been in its effects on choice, but on acquisition and its asymptote. Although Hull (1943) used the data from choice situations, such as Perin's (1943) and Anderson's (1932), he treated them as if they revealed only the habit strength of the response whose reinforcement was being delayed and not the operation of a choice mechanism. From these data, Hull concluded that the asymptotic habit strength of a response was a negative exponential function of the delay of reinforcement. At least superficially, Chung's study (1965b) confirms Hull's hypothesis. As stated earlier, Chung studied, in a two-choice procedure for pigeons, the effects of delaying reinforcement for one of the two alternative responses, while the other alternative was always reinforced immediately. As the duration was varied it was found that the effects could be adequately summarized by a decreasing exponential function between the relative frequency of responding on the key for which reinforcement was delayed and the duration of the delay. This agreement with Hull is all the more impressive for having been based on results from different species—pigeons versus rats—and from a number of different types of apparatus, from the simple T-maze to the two-key pigeon box. The agreement notwithstanding, certain problems remain unsolved.

There is, first, the fact that in picking the exponential function Hull was relying heavily on Anderson's results, obtained from an old-style discrimination-box procedure in which rats chose one of either of two or four compartments presented simultaneously. Any choice was rewarded with food, but, depending upon the compartment chosen, the rat was detained for periods of 1 to 4 min. A rereading of Anderson's original paper shows that although it is clear that at some point in training the rats were distributing their choices in approximate agreement with an exponential function, his data suggest that with further training the rats were tending to choose only the compartment with the shortest delay. Such a tendency is hardly surprising, for there is nothing in the situation to dispose a rat to choose a longer delay when a shorter one is available and equally profitable. That situation existed in Anderson's study, but not in the present experiment, as is shown below. In his curve-fitting, Hull used only the data from the earlier point in training, and spoke of the distribution of choices as if it were dictated by the rats' inability to form discriminations between certain short delays and longer ones, thereby tacitly making the effect of delay one of discriminability rather than of reinforcing power.

Whatever the answer may be as regards Anderson's study and the problem of temporal discrimination in the rat, the parallel to Chung's study is obviously questionable. Since the later study used a pair of variable-interval schedules, with delays of reinforcement for one of the choices, there was a factor disposing the pigeons to respond to both alternatives. Unlike the continuous reinforcement in Anderson's experiment, and, incidentally, in virtually all of the other experiments in the preoperant literature, the aperiodic schedule may impose a penalty in reinforcements lost should the animal respond exclusively to either alternative, the one with the shorter delay or otherwise. Chung's exponential, then, may well have been describing an asymptotic performance, but its agreement with Anderson's preasymptotic data may be devoid of substantive meaning.

This is not the only problem. Hull noted that a number of experimenters had examined the effects of reinforcement delay and had, contrary to their own expectations, found virtually none for delays that, in other studies, proved to be more than ample. John B. Watson was one of these, having varied delay of reinforcement in his version of the puzzle box (1917) and found that his rats were unaffected by delays up to 30 sec. More recently, Ferster (1953) argued that delays of even longer durations can be bridged with no decrement in performance if the animal is equipped with the suitable behavior to mediate the temporal gap. Ferster was, in effect, agreeing with Hull, who also attributed the lack of potency of delays to the interfering effects of secondary reinforcement. And more recently still, Logan (1960) examined reinforcement delay in a runway for rats and found relatively minor quantitative effects of delays between 1 and 30 sec.

Finally, the present study, in which the pigeons were choosing between pairs of delays, suggests, as will be shown, that the exponential function may not be the best summary of the effects of delay, the earlier agreement notwithstanding. The data in Figure 5.2 are replotted in Figure 5.6 to test the validity of the negative exponential function as a description of the results. The ordinate is the relative frequency of responding on a logarithmic scale; the abscissa is the duration of the delay on a linear scale. Negative exponentials plot as straight lines with negative slope in such a semi-logarithmic coordinate. The three straight lines shown were

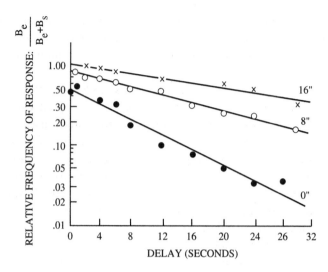

Figure 5.6   Relative frequency of responding on the experimental key, on a loga-rithmic scale, as a function of the delay for that key, averaged across subjects. The parameters refer to the duration of the standard delay.

drawn by eye through the data points of the present experiment and of Chung's earlier study. The parameters on the curves give the duration of the standard delay for each of the three conditions.

In the absence of other considerations, the fit between data and theory in Figure 5.6 would undoubtedly enhance the credibility of both. Except for a slight tendency toward upward concavity, most evident for the bottom function, the points hover close to the straight lines called for by the hypothesis of a negative exponential. The data used by earlier workers to substantiate the negative exponential rarely fit so well, over so broad a range, with so little averaging necessary. The other consideration, however, that makes the negative exponential suspect, is the function in Figure 5.3, which shows that the relative frequency of responding matches the complement of the relative duration of the delay. It is readily shown that this matching relation is mathematically incompatible with the negative exponential suggested by Figure 5.6. No proof will be given since it is virtually self-evident that the two following equations, representing the matching rule and the exponential rule, respectively, cannot both hold:

$$\frac{B_e}{B_e + B_s} = \frac{D_s}{D_e + D_s}, \tag{5.3}$$

$$\frac{B_e}{B_e + B_s} = ka^{-bD_e}, \qquad\qquad (5.4)$$

in which k, a, and b are constants and the other symbols are as used above.

Two incompatible theories can fit a single set of data points only if there is enough variability in the data to provide the requisite level of ambiguity. The present data clearly provide at least this level. Fortunately, however, the quantitative nature of the two theories permits one further step in the analysis. If Equation 5.3 did, in fact, correctly describe the data, the points would deviate from the straight lines in Figure 5.6 in tending to be concave upward. Moreover, the upward concavity would be greatest for the bottom curve and so slight as to be virtually undetectable for the upper curve, with the middle one falling between. This mathematical implication of the matching function seems to be borne out by the pattern of deviations of the data from the straight lines. On the other hand, the contrary assumption, that Equation 5.4 is correct, predicts that the points in Figure 5.3 would tend to deviate from the diagonal by being concave downward, with the discrepancy between predicted and obtained values increasing at both ends of the function. This suggestion appears to account for none of the variance of the points around the theoretical line.

The present experiment was not designed as a test of the two formulations being discussed. It is not surprising, then, that a firm conclusion is not forthcoming. The data slightly favor the matching relation, but not enough to exclude the exponential. On the other hand, it is now clear that the two theories, in spite of their mutual incompatibility, predict similar relative frequencies of responding within the range of delays usually studied. To separate between the theories quantitatively would require an experiment in which very long and very short delays would be directly compared in a choice procedure; it is at these extremes that the two theories diverge measurably.

There are, however, non-quantitative arguments favoring the matching rule. The theoretical curve in Figure 5.3, which asserts that a given ratio of delays will produce a given ratio of responses, independent of the absolute levels involved, is the sole theoretical line allowed by the matching hypothesis. In contrast, the exponential hypothesis allows any negative slope and intercept in the semi-logarithmic coordinate used in Figure 5.6. The matching hypothesis, in other words, is by far the more restrictive and powerful of the two. This is also evident from the fact that Equation 5.3 calls for no free parameters, while Equation 5.4 calls effectively for two.

The approximate symmetry of the rising and falling curves in Figure 5.5 means that the number of pecks on the two keys summed together maintained an approximate constancy, in spite of the changing duration of delay for one of the keys. A comparable insensitivity of responding to the duration of delay was already noted in connection with experiments by Watson (1917), Ferster (1953), and Logan (1960). The results of these experiments, and the summed rate of responding in the present experiment, have one thing in common: all are based on absolute measures of performance. The results obtained by Anderson (1932), Perin (1943), and Chung (1965b), on the other hand, are based on relative measures. In the few studies that show large effects of delay and that use absolute measures, e.g., Skinner (1938, p. 139ff.) or Dews (1960), the animal's response postpones reinforcement. This is a direct contingency favoring a low rate of responding and not reinforcement delay in the usual sense of the term. Delay of reinforcement, then, like frequency of reinforcement (see Chapter 1) or amount of reinforcement (Catania, 1963a), is a variable that influences more the relative strength of a response among a set of response alternatives than the absolute strength of a response in isolation. Given this insensitivity of the overall rate of responding in the present experiment, the various features of Figure 5.5 follow from the matching relation depicted in Figure 5.3. It then follows, for example, that a change in delay duration on one key will alter in opposite directions the rates of responding on both keys, and that the functions become progressively less curved with increasing standard delays. For all its simplicity, the matching hypothesis appears able to account for the present results, and also for those to be found in the experimental literature.

6

Self-Control as Response Strength

We call behavior impulsive when it might not have occurred at all if its long-range consequences had been given full weight. We eat dessert impulsively, for example, if we had been planning to say no but at the last moment yielded to the temptations of the pastry cart. Self-control is the other side of the coin. If we decline the pastry, in spite of a sense of pleasure missed now so as to avoid regrets later, we may say we controlled ourselves. Self-control may seem paradoxical in a deterministic system, but it is not. In a reinforcement account of behavior, self-control is itself controlled by contingencies of reinforcement, just as controlling another organism is. For self-control, the critical contingencies depend on delay of reinforcement.

In impulsiveness versus self-control, time is always of the essence. Behavior is impulsive if it sacrifices long-range considerations for short-range gains; it is self-controlling if the reverse is the case. There may seem to be additional distinctions between impulsiveness and self-control. Impulsive acts typically have highly localized sensory or emotional consequences; self-control often produces more diffuse and less hedonistic results. For example, delicious tastes or sexual adventures compete with the more tranquil pleasures of good health, fitting into one's clothes, or keeping one's family intact. These distinctions can, however, be traced back to the temporal displacement of the competing alternatives. The

Originally published in C. M. Bradshaw, E. Szabadi, and C. F. Lowe (eds.), *Recent Developments in the Quantification of Steady-State Operant Behavior* (Amsterdam: Elsevier/North Holland Biomedical Press, 1981), pp. 3–20. Preparation of this essay was supported by Grants MH-15494 from the National Institute of Mental Health and DA-02350 from the National Institutes of Health to Harvard University, hereby gratefully acknowledged. The author also wishes to express appreciation to Robert Boakes, Stephen Lea, and Mark Snyderman for helpful comments.

quicker reinforcer is necessarily localized in time, otherwise it could not be truly quick, which implies mainly discrete sensory and emotional stimuli. In contrast, the long-range consequences of behavior can spread over substantial intervals, and hence will include such non-sensory abstractions as healthy living or family harmony, states that entail integration over time.

Ainslie (1974, 1975), Rachlin (1970), and Rachlin and Green (1972) first recognized that hyperbolic delay-of-reinforcement gradients could be a clue to the psychology of self-control. Each of them made use of the experimental finding (see Chapter 5) showing the relative frequency of responding matching the relative immediacy of its reinforcement, immediacy defined as the reciprocal of delay. This finding suggested that delay (or immediacy) is a dimension of reinforcement no different in principle from amount or rate of reinforcement. Like the other dimensions, delay (D_i) was approximated by the matching law for two alternatives, B_1 and B_2:

$$B_1/B_2 = D_2/D_1. \tag{6.1}$$

Ainslie and Rachlin both noted that the hyperbolic relationship between delay and reinforcing value implicit in Equation 6.1 clarified the familiar collapse of self-control in the face of temptation. The collapse of self-control has come to be called preference reversal and is usually illustrated as in Figure 6.1. A subject confronts a pair of mutually exclusive alternatives. The larger reinforcement is more delayed than the smaller. Because the delay functions are hyperbolic, they must cross. Because they cross, it follows that at longer delays, the later reinforcer will seem more valuable and at shorter delays, the earlier one will.

The curves in Figure 6.1 trace two hyperbolic equations. The earlier alternative has a reinforcer of fixed amount A_1 delayed by a variable duration, D, which is given by $(T - t)$ for the earlier alternative and $[(T + 3) - t]$ for the later (see caption to figure). In general, the later alternative's reinforcer, A_2, is aA_1 in which $a > 1$, and its delay is $D + \Delta$ in which $\Delta > 0$. The multiplier a is the ratio of the two reinforcing values independent of delay.[1] Taking delay into account, Chung and

---

1. In this article Herrnstein uses a to stand for either A_2/A_1, the ratio of the amounts of the individual reinforcers, or R_1/R_2, the ratio of their rates over a common session duration. [Editors' Note]

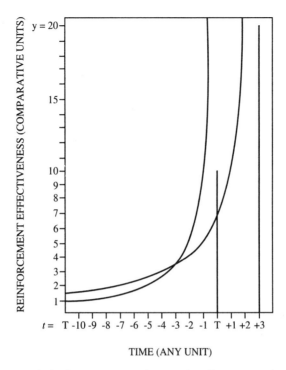

Figure 6.1   Hyperbolic discount curves showing the effectiveness of two hypothetical reinforcements as a function of delay: the effectiveness, at any given moment of the time axis, T, of a reinforcement occurring at time T versus a reinforcement twice as large occurring at time $T + 3$. Left curve shows $y = 10/(T - t)$; right curve, $y = 20/[(T + 3) - t]$. (Figure and caption from Ainslie and Herrnstein, 1981.)

Herrnstein's finding (see Chapter 5) suggests that the reinforcing values, V_i, are as follows:

$$V_1 = A_1/D, \qquad\qquad\qquad\qquad\qquad\qquad (6.2a)$$

$$V_2 = aA_1/(D + \Delta). \qquad\qquad\qquad\qquad\qquad (6.2b)$$

The preference-reversal point (D_p) would be where $V_1 = V_2$ and is readily derived from Equations 6.2a and 6.2b:

$$D_p = \Delta/(a - 1). \qquad\qquad\qquad\qquad\qquad\quad (6.3)$$

Since D_p measures the time to the earlier alternative, the larger it is, the greater the period during which the impulsive alternative is prepotent over the forbearing alternative. D_p therefore estimates the overall strength or likelihood of the impulsive alternative. Equation 6.3 shows that D_p is directly related to Δ, the temporal displacement of the later reinforcer, and inversely related to a, the ratio of the later reinforcer to the earlier, notwithstanding delay. Both of these implications satisfy common sense about self-control, as does the effect of letting D increase without limit in Equations 6.2a and 6.2b. Consider the ratio V_2/V_1 as D approaches infinity:

$$\frac{V_2}{V_1} = \lim_{D \to \infty} \frac{\frac{aA_1}{D+\Delta}}{\frac{A_1}{D}} = a. \tag{6.4}$$

At a sufficient perspective in time, the ratio of reinforcing values approaches a, which is in fact what they are without regard to delay.

In spite of its face validity and appealing simplicity, this formulation of self-control seems to me to be conceptually inadequate for the three reasons given below. Before listing the reasons, I note that they are conceptual; empirical results will be considered later.

Criticisms of the Ainslie-Rachlin Theory

1. Figure 6.1 has as its ordinate "reinforcement effectiveness." Ainslie typically (e.g., 1975) refers to this as "reward effectiveness"; Rachlin (1970) as "value of reinforcement"; Brown and Herrnstein (1975) as "value of reward." Although these essentially equivalent expressions are consistent with Chung and Herrnstein's findings, they are meant to say something about the occurrence of behavior. That is to say, there is a link missing, a link to connect behavior with reinforcing value. Ainslie and Rachlin both imply a monotonic relationship between reinforcing value and behavioral strength, and leave it at that. The connection between reinforcement delay and behavior presumably is compatible with the matching law, as Chung and Herrnstein's results suggest it must be to a first approximation, but it remains to be shown explicitly how. If reinforcing value decreases hyperbolically with reinforcement delay, what can we say about response strength as a function of reinforcement delay? Later, I will offer an answer.

2. The hyperbolic form of Equation 6.2 says that reinforcing value approaches infinity as delay of reinforcement approaches zero. Conse-

quently, for all temporally displaced choices in which the later alternative has the larger reinforcer, there must be a crossover or preference reversal. Appearances to the contrary, this does not necessarily imply the patently false conclusion that self-control always collapses, for two reasons. First, subjects confronted with temporally displaced alternatives may precommit themselves irrevocably to the later alternative at delays greater than D_p. If dieters could order the day's meals the night before and have no other access to food, it would probably be easy to lose weight. A habitual gambler may cease going to race tracks but still find placing a bet irresistible when he gets too close to an off-track betting parlor. Not going to race tracks is a form of precommitment against gambling, but in ordinary environments, irrevocable precommitments are hard to arrange. Second, when D_p is smaller than the duration of the response required for the earlier reinforcer, then there should be no preference reversal because the impulsive response would have to start at delays greater than D_p, at which times the self-controlling alternative is prepotent. Nevertheless, it seems counter-intuitive (at least to me) that, except for these two conditions, all temporally displaced alternatives involve preference reversal.

3. The Ainslie-Rachlin formulation lacks a parameter for expressing variations in time-discounting rates. For qualitative discussions of self-control, this is not so much a deficiency as it is a conventional simplification. However, variation in self-control is a central concern in its own right. A comprehensive theory of impulsiveness and self-control needs to say something about individual, species, and procedural differences. Pigeons and other sub-human subjects fail to forbear for delays of a few seconds in access to food, whereas under some conditions human subjects tolerate delays of days, months, or years. The differences are so great that it may strain credibility to say that a single framework can encompass the animal and human extremes. The model presented below presumes to do just that.

A Response-Strength Theory of Impulsive and Forbearing Behavior

A theory should correct the deficiencies of the earlier theory while preserving its advantages. An improved theory should: (a) predict preference reversal under some, but not all, conditions; (b) give a formal representation of variations in time discounting; (c) explicitly predict behavior itself, rather than just the inferred value of reinforcement. Three further considerations seem reasonable for a theory of impulsive behavior, given existing data. A theory should: (d) be consistent with matching between relative response and relative immediacy (Equation 6.1); (e) converge on the normative matching law (given below as Equation 6.5) as delay of reinforcement approaches zero; (f) suggest how an organism's current psychological state and past history interact with self-control.

The normative matching law states that behavior, B_1, is a hyperbolic function of its reinforcement, R_1, as follows:

$$B_1 = kR_1/(R_1 + R_e). \tag{6.5}$$

B_1 and R_1 are usually measured as rates of occurrence, but may be thought of more generally as strengths of behavior and reinforcement during a standard condition of observation. The remaining terms, k and R_e, are commonly just curve-fitting parameters but can be thought of as measures, respectively, of total behavior, including B_1 and all competitors and of reinforcement for the competitors. At least approximately, Equation 6.5 fits most data on asymptotic reinforced responding (see Chapter 2). Now we want to add the variable, D, to Equation 6.5 so as to meet the six conditions listed above.

The simplest solution[2] I have found is to introduce a parameter, I.

---

2. Other solutions may seem simpler, but the ones I have considered fail to do the job. In particular, here are three discarded alternative formulations:

(a) $B_1 = \dfrac{\frac{kR_1}{D}}{\frac{R_1}{D} + \frac{R_e}{D}}$.

The problem here is obviously that D cancels out, leaving B_1 unaffected by delay.

(b) $B_1 = \dfrac{\frac{kR_1}{D}}{\frac{R_1}{D} + R_e}$.

Here the problem is that B_1 approaches k as D approaches zero. Not all immediately reinforced responses are at asymptote, as far as we know.

(c) $B_1 = \dfrac{k\frac{R_1}{D}}{\frac{R_1}{D} + \frac{R_e}{D} + R_e} = \dfrac{kR_1}{R_1 + R_e(D + 1)}$.

This shares several features with the chosen equation (6.6), but has the disadvantage of combining into a single parameter, R_e, what seems more appropriately segregated into two parameters, R_e and I in Equation 6.6.

$$B_1 = \frac{\frac{kR_1}{D}}{\frac{R_1}{D} + \frac{R_e}{D} + I} = \frac{kR_1}{R_1 + R_e + DI}. \tag{6.6}$$

The new parameter is, to begin with, a scale factor for delay (D). Subjects that discount time sharply have high values of I, and, as will be shown later, appear impulsive, and vice versa for gradual discounters. Equation 6.6 retains the relationship between reinforcement value and reinforcement delay of the Ainslie-Rachlin theory, but embeds it in a response-strength formula modeled on the matching law. When delay is absent, Equation 6.6 reverts to normative matching, i.e., Equation 6.5.

Having described Equation 6.6 to other audiences, I know that it is sometimes disapproved because it introduces a new parameter. It may strike some people as a step away from the data, hence away from theoretical parsimony. However, a new parameter for delay seems unavoidable because of individual, species, and procedural differences. Delay, measured in the physical units of time, must be buffered through some sort of parametric transformation in its relation to behavior; no transformation is simpler than the one in Equation 6.6. The precedent for adding parameters is already implicit in the normative matching law itself: k is the parameter for transforming measurements of behavior and R_e is for transforming measurements of reinforcement.

In standard concurrent procedures, the parameters vanish for predictions of relative responding. Consider Equation 6.5 applied to a pair of concurrent variable-interval schedules, with equivalent response topographies, and with B_1 and B_2 reinforced by two frequencies of the same reinforcer, R_1 and R_2, respectively:

$$B_1 = \frac{kR_1}{R_1 + R_2 + R_e}, \qquad B_2 = \frac{kR_2}{R_1 + R_2 + R_e};$$

$$\frac{B_1}{B_2} = \frac{R_1}{R_2}. \tag{6.7}$$

Equation 6.7, which is the ratio form of the matching law, deals only with the directly observable measures of responding and reinforcement, the transforming parameters having been cancelled out. It is no doubt because the parameters disappear that the matching law was stated first in its Equation 6.7 form, rather than in its Equation 6.5 form. It is easier to fit data to a known equation with free parameters, but, for that very reason, harder to discover it in the data for the first time.

Reinforcement delay is analogous. Chung and Herrnstein varied delay for a pair of alternatives reinforced on two otherwise concurrent variable-interval schedules. Applying Equation 6.6 to each alternative, and bearing in mind that on conc VI VI, each alternative's reinforcement is represented[3] in both denominators:

$$B_1 = \frac{k\frac{R}{D_1}}{\frac{R}{D_1} + \frac{R}{D_2} + \frac{R_e}{D_1} + \frac{R_e}{D_2} + I}, \qquad B_2 = \frac{k\frac{R}{D_2}}{\frac{R}{D_1} + \frac{R}{D_2} + \frac{R_e}{D_1} + \frac{R_e}{D_2} + I};$$

$$\frac{B_1}{B_2} = \frac{D_2}{D_1}. \tag{6.8}$$

Equation 6.8 repeats Equation 6.1, which was approximated by Chung and Herrnstein's results. As before, the matching law for symmetric choice cancels out the parameters, including I in this instance.

Plotting B_1 as a function of D in Equation 6.6 produces something close to the familiar upwardly concave hyperbolas of the Ainslie-Rachlin theory, except that the ordinate measures observable responding rather than inferred reinforcing value. Figure 6.2 illustrates Equation 6.6, with a few values of I. As I decreases, behavior becomes less rapidly weakened by delay, in the limit becoming totally unaffected by it. The level of behavior reached when $D = 0$ (immediate reinforcement) depends on the simple matching law, i.e., on k, R_1, and R_e. The present formulation says how these other factors interact with delay.

Before discussing the interactions, we should consider preference reversal from the perspective of Equation 6.6. At issue are a pair of mutually exclusive alternatives, B_1 and B_2. For setting up our equations, it is important that the alternatives are mutually exclusive, hence not strictly comparable to conc VI VI, for which we assumed that the subject was receiving all the reinforcements from both alternatives. As in Equations 6.2a and 6.2b, B_1's reinforcement is R_1 and delay is D, whereas B_2's reinforcement is aR_1 and its delay is $D + \Delta$

$$(a > 1; \Delta > 0).$$

3. In a true concurrent schedule the alternatives must have the same denominators, which is why the derivation of Equation 6.8 proceeds as it does. In fact, delay of reinforcement is problematic in this respect. When the subject is reinforced by one alternative, reinforcement from the other alternative is precluded for the duration of the delay, possibly a substantial fraction of the interreinforcement interval. Thus, Chung and Herrnstein's procedure is more nearly a concurrent-chain than a true concurrent, which may account for the deviations from matching in the data noted by Williams and Fantino (1978). Later, when I discuss Green and Snyderman's (1981) findings, the same issue recurs.

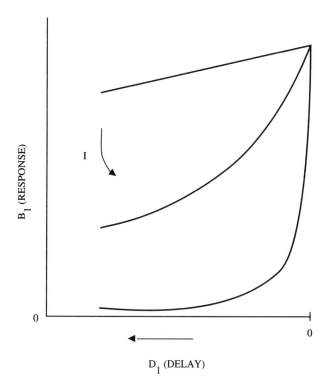

Figure 6.2   Plotting Equation 6.6 with reinforcement delay (D_1) varying along the x-axis. Each curve is for a different value of I; the other quantities in Equation 6.6 were held constant.

Applying Equation 6.6 to each alternative in the simplest[4] case:

$$B_1 = kR_1/(R_1 + R_e + DI); \qquad B_2 = kaR_1/[aR_1 + R_e + (D + \Delta)I].$$

Each alternative's reinforcer does not appear in the other's denominator because they are mutually exclusive. Preference reverses when the curves pass through the point $B_1 = B_2$, illustrated in Figure 6.3 at a delay of D'_p. The value of D'_p is given by:

$$D'_p = \Delta/(a - 1) - R_e/I. \tag{6.9}$$

---

4. It is the simplest case because the same parameter values are assumed for each alternative. In principle, each of the parameters may differ.

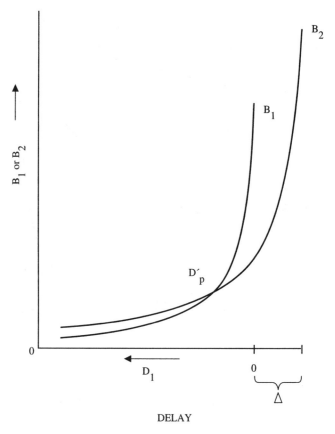

Figure 6.3   Plotting Equation 6.6 for a pair of alternatives, B_1 and B_2. B_1's reinforcement is delayed by D_1; B_2's, by $D_1 + \Delta$. B_1's reinforcement is R_1; B_2's, aR_1. The delay of B_1 at preference reversal, D'_p, is given by Equation 6.9.

The first term on the right in Equation 6.9 is the same as the controlling quantity in the Ainslie-Rachlin account of preference reversal (see Equation 6.3). This term contains the self-evident controlling variables, namely, the temporal displacement of the alternatives (Δ) and the ratio of their reinforcing values exclusive of delay (a). Comparing Equations 6.3 and 6.9 shows that the response-strength theory contains and extends the Ainslie-Rachlin theory. The second term in Equation 6.9 adds two contextual variables, which plausibly relate to self-control.

The numerator of the second term, R_e, expresses the overall context of reinforcement vis-à-vis the reinforcement contained in R_1 and R_2. If the context is rich in other alternatives, or if R_1 and R_2 are feeble reinforcers because of minimal incentive or drive, then R_e is large and D'_p is small. That is to say, self-control is directly associated with the strength of the reinforcing context outside the two alternatives in question.

In most self-control situations, different drives and reinforcers are involved in R_1 and R_2. For example, R_1 may be the reinforcing effect of a fattening dessert and R_2 is the avoidance of a weight gain; R_1 may be smoking cigarettes and R_2, the avoidance of the long-term risks. For such cases, changes in drive or incentive are typically asymmetrically related to the two alternatives. If dessert arrives while one is still hungry, impulsiveness is more likely not just because R_e is smaller than when one is not hungry but also because a is smaller too. When the drive applies equally to both alternatives, then the effect on choice would be entirely due to the change in R_e, since a would be unchanged, but such cases do not readily model self-control as usually conceived, with motivational states pitted against each other. Nevertheless, an example of changing self-control within a single drive would be when choosing at, say, 11 A.M. between lunch and supper, assuming that only one can be eaten and that supper is more reinforcing, aside from the effects of greater delay. Equation 6.9 says that, the hungrier you are, the more likely you are to choose lunch, because R_e varies inversely with hunger. The Ainslie-Rachlin model, which omits R_e, predicts that choice between lunch and supper would be independent of hunger.

The other quantity in the right term, I, is the rate of time discounting, which may depend on various influences besides individual and species differences. Variation may also be situational, for example depending on whether stimuli are provided to fill the gap between response and reinforcer. It may vary from reinforcer to reinforcer, or from response topography to response topography. It may depend on past learning. In short, the new parameter modulates the effect of time as if it were a scale factor and in doing so may capture the effects of numerous influences on self-control. Its conceptual status is suggested by its dimensions.

Dimensional analysis of Equation 6.9 shows that I must be measured in units of reinforcement per unit time. That is to say, its units are the same as R_e's, but then divided by time. Consider again the fundamental expression (Equation 6.6) for the effects of delayed reinforcement:

$$B_1 = kR_1/(R_1 + R_e + DI).$$

The term DI affects B_1 the same way as R_e, inasmuch as it quantifies reinforcement in competition with R_1, the reinforcement for B_1. To preserve dimensional balance, DI must have the same dimensions as R_e; therefore it follows that I taken alone must have the dimensions of R_e divided by time. Substantively, this means that the competing reinforcers come in two forms: R_e is the competing reinforcement in the immediate temporal vicinity of B_1 and is therefore not dependent on delay; DI is the competing reinforcement in the interval between the response and its eventual reinforcement and is therefore strictly dependent on delay. The term DI expresses how, from the subject's point of view, the case against emitting B_1 grows with delay. A host of factors can influence the case one way or another, and they are all funnelled willy nilly into a single multiplicative parameter, I. This may seem too simple for so complex a matter, but it is meant to be a point of departure for an issue that has simply been omitted in previous models.

When R_e/I equals or exceeds $\Delta/(a - 1)$, D'_p falls to zero and beyond into "negative" delays (see Figure 6.3). This means that preference does not reverse at all for certain ranges of parameter values, namely when

$$\Delta/(a - 1) \leq R_e/I. \tag{6.10}$$

Equation 6.10 states the relationship among the parameters defining a dividing line between self-control and impulsiveness. Qualitatively, the equation fits common sense, although it may be hard to discern how without further clarification. We promote self-control by reducing the delay to the later, larger reinforcer (Δ) or the scale factor on delay (I), or by increasing the ratio of the reinforcing values of the later to the earlier choice (a) or the reinforcing context exclusive of the two choices (R_e).

Figure 6.4 attempts to visualize Equation 6.10. The highest value of a (ratio of later to earlier reinforcement, by definition greater than 1.0) for which preference reverses is shown as a function of the displacement of the later reinforcement (Δ). The function is a line with slope equal to I/R_e. As long as a is at or below the function, preference reverses at some value of D (given by Equation 6.9). Increasing values of Δ relax the limit on a. Above the limit, preference will not reverse, and the subject will choose the later, larger reinforcer at all delays, thereby controlling himself. However, this relationship between a and Δ is modulated by the background variables I and R_e, specifically, by their ratio. The smaller the ratio I/R_e, the lower the limit on preference-reversing values of a, hence the more resistant is the subject to impulsiveness. Figure 6.4 thus demonstrates the four critical variables interacting for preference reversal, according to the present theory. According to the Ainslie-Rachlin theory, there is only one function on these coordinates, a line of infinite

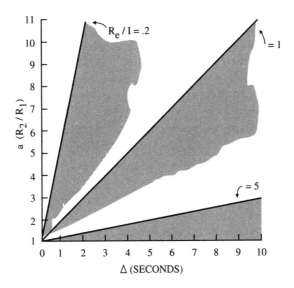

Figure 6.4   For a given ratio, R_e/I, preference reverses if a is at or below the indicated line. This illustrates the relationship in Equation 6.10.

slope rising from $a = 1$ and $\Delta = 0$. That is to say, their model says that for any ratio of R_2/R_1 greater than 1.0, with any displacement of R_2 greater than 0, preference reverses except for the two conditions noted earlier.

Evidence.   Two sorts of evidence are germane to the present analysis of self-control. First is the question of preference reversal itself. Although preference reversal does not distinguish between the Ainslie-Rachlin model and the present modification of it, it is corollary to any hyperbolic theory of impulsiveness. There would be no point in distinguishing among hyperbolic theories unless there was a good reason, such as preference reversal, to select among them to begin with. Preference reversal can be taken as having been demonstrated, although not unambiguously in every test (Ainslie, 1974; Deluty, 1978; Fantino, 1966; Logan and Spanier, 1970; Navarick and Fantino, 1976; Rachlin and Green, 1970; Solnick et al., 1980). These studies report a variety of examples ranging from clear to questionable, using rats, pigeons, and humans as subjects. I'll describe the experiment on humans (Solnick et al., 1980) as representative of the lot.

Volunteer subjects in an introductory psychology class thought they were participating in a study of the effects of noise pollution on performance at mathematics problems. While working on their problems, they were periodically subjected to a loud white noise which could be turned off for a minute or so by pressing either of two buttons. One button shut off the noise sooner than the other, but for a shorter silent period. Preference reversed as a function of delay: a clear majority chose the short-early silent period when it came immediately but the large majority switched to the other alternative when a 15-second delay was added to both alternatives.

To distinguish between the Ainslie-Rachlin and response-strength theories, we need to look at the details of preference reversal more critically, for the theories diverge in their fine structure. Both theories make predictions about the relative strengths of two alternatives at any delay. The formal expressions given earlier yield the limiting values shown in Table 6.1. For either model, the ratio of alternative strengths rises towards a, the ratio of the larger to the smaller reinforcer, as the delay approaches infinity. Thus, the ratio would show undermatching at all finite delays. Deluty (1978) and Navarick and Fantino (1976), studying delays of punishment and reinforcement in a conventional concurrent-chain procedure, found persistent undermatching, as both theories imply. However, Navarick and Fantino found that when two reinforcers were equally delayed (i.e., $\Delta = 0$ in the present model), increasing the delay increased the preference for the larger reinforcer towards a and beyond. In the Ainslie-Rachlin model, the ratio of alternative strengths should be a at all delays if the delays are equal (see Equation 6.4); in the response-strength model, the ratio of alternative strengths with equal but finite delays should undermatch, and the more so the shorter the delay (see Table 6.1 with $\Delta = 0$). The Navarick and Fantino finding therefore favors the present theory, except that preference became more extreme than matching at long delays. However, for mutually exclusive alternatives, extreme preference is expected because of the feedback between choice and reinforcement frequency. This point recurs when we consider Green and Snyderman's data below.

Table 6.1   Limiting values of B_2/B_1

	Ainslie/Rachlin	Response strength
$\lim_{D \to 0} \dfrac{B_2}{B_1}$	0	$a\left(\dfrac{R_1 + R_e}{aR_1 + R_e + \Delta I}\right)$
$\lim_{D \to \infty} \dfrac{B_2}{B_1}$	a	a

Table 6.1 also provides a clue to the conditions for preference reversal. For each column, preference reverses if 1.0 falls in the interval between its upper and lower quantities. For Ainslie-Rachlin, this is always the case (since $a > 1.0$). For the other theory, it depends on the parameters, as plotted in Figure 6.4. Mazur and Logue (1978) presented evidence that deliberate training can reduce or eliminate preference reversal. They showed that pigeons can learn to choose the later, larger reinforcer if the value of Δ gradually increases. The subjects in their study chose 6 seconds of food delayed by 6 seconds over 2 seconds of food delivered immediately, but only after the later reinforcement was gradually displaced. Without the special training, they consistently chose the small-early reinforcement. We can interpret the special training as having reduced the value of I. Studies on unsignalled reinforcement, on the other hand, show very sharp discounting with brief delays (e.g., Richards and Hittesdorf, 1978; Sizemore and Lattal, 1978; Williams, 1976), as if I is relatively large when no stimulus ties the response to its delayed reinforcement.

The results of one other experiment may also show the effects of a changing value of I. Christensen-Szalanski et al. (1980) gave thirsty rats a choice between a small dip of water presented earlier and a larger dip, presented later. For any given pair of delays, they found preference shifted toward the later alternative as water deprivation was increased. Taken alone, this finding violates Equation 6.9, which says that impulsiveness should increase as R_e falls. However, Christensen-Szalanski et al. also showed an equal or larger shift toward the later alternative as a result of a past history of water deprivation. The effect of past history was taken as evidence that the rats had learned strategies to mediate long-range consequences of behavior with respect to water needs. In present terms, this means a drop in I, which would shift preference toward the large-late alternative. From the data presented, it was not possible to distinguish between the effects of current and past deprivation, which suggests that the change in I (which would presumably affect both treatments) dominates the change in R_e (which would affect only the current deprivation treatment).

A more direct assessment of delay theories is afforded by Green and Snyderman's experiment. Using concurrent chains, they studied the effects of delay duration on choice. The ratio of reinforcer amounts was 3 in each of three procedures associated with a particular ratio of the delays, namely, $3:1$, $6:1$, and $3:2$. The absolute durations of delay ranged from 6 to 12 seconds to 120 seconds (for the large-late alternative) in each procedure. The response-strength theory accounts for the results qualitatively: preference for the large-late alternative declines with the delay for the $3:1$ and $6:1$ conditions, but rises for the $3:2$ condition. These inferences follow from the following expression derived from Equation 6.6, when the suitable proportionalities are inserted $[a = R_2/R_1; T = (D + \Delta)/D]$:

$$\frac{B_2}{B_1} = \frac{a(R_1 + R_e + DI)}{aR_1 + R_e + TDI}. \tag{6.11}$$

The ratio, a, was 3 in all cases. The three procedures varied T, as shown in Table 6.2.

The theory says that B_2/B_1 must fall with increasing delay for procedures 1 and 2, but it may or may not rise for procedure 3, depending on the relative sizes of a, R_1 and R_e. Green and Snyderman noted that neither the Ainslie-Rachlin theory nor Navarick and Fantino's (1976) equation for concurrent chains handled their data, but the response-strength formula was not available to them for evaluation.

Had it been available, they would surely have raised the question of quantitative, rather than merely qualitative, agreement. The observed ratios of B_2/B_1 tended to be more extreme than the ones predicted by the response-strength theory at the longer delays. This discrepancy highlights a weakness in the theory, but it may be a weakness of omission rather than of commission. The theory as formulated here is for mutually exclusive choices with mutually independent consequences. A concurrent chain procedure almost invariably involves interactions between the choices. The first-link schedule may, for example, save reinforcers, so that the local rate of reinforcement for one alternative varies inversely with its recent frequency of having been chosen. In addition, the theory as here expressed assumes that a is fixed, independent of the pattern of choices. In a concurrent chain, reinforcement rate interacts with choice so as to create feedback with the value of a. Whether the response-strength theory, suitably elaborated with feedback functions, can account for asymptotic choice in concurrent chains remains to be tested,

Table 6.2   Predicted limiting values of B_2/B_1

	As $D \to 0$	As $D \to \infty$
Procedure 1. ($T = 3$).	> 1.0	1.0
Procedure 2. ($T = 6$).	> 1.0	0.17
Procedure 3. ($T = 3/2$).	> 1.0	2.0

but even in its present, simplest form it appears to be a qualitative step forward.

Conclusion

Six requirements were considered for the response-strength theory of self-control, as repeated here and further specified. (a) The theory embeds hyperbolic delay-of-reinforcement gradients in the matching equation, with the result that preference reverses or not depending on the ratio of reinforcer values, the temporal displacement of the late-large reinforcer, a time-discounting parameter, and the overall context of reinforcement (see Figure 6.4). (b) Individual, species, and procedural differences depend on a parameter that is, in effect, a scale factor for duration of delay. (c) The matching equation for absolute strength of responding connects inferred reinforcing value to behavior. (d) For truly concurrent alternatives, the theory predicts matching between relative responding and relative immediacy. (e) When delay is zero, the theory reverts to the normative matching equation (Equation 6.5). (f) Current psychological state and past history are taken account of by parameters for overall context of reinforcement and for time-discounting rate.

Embedding the Ainslie-Rachlin formula for the effect of delay of reinforcement in a version of the matching equation is not a very large step conceptually. Even so, the two theories diverge over a surprisingly large range of conditions, with the balance of the minimal data in hand favoring the present one. However, neither theory has been tested quantitatively to any extent. We can say that a hyperbolic response-strength formulation of self-control seems promising, that it agrees with existing data at least qualitatively, and that it suggests a variety of quantitative tests.

A hyperbolic process has evolutionary implications (see also Ainslie and Herrnstein, 1981). The crossing curves mean that temporal perspective toward the alternatives approaches the relative valuation they would have without delay. As the alternatives draw near, there is a growing tendency to overvalue the earlier reinforcer. Since delayed consequences in the real world usually are more uncertain than immediate pay-offs, a bias toward impulsiveness seems to be the adaptive *initial* position in any environment. Also, if creatures were not thus biased, they would be faced with a formidable challenge in bookkeeping, trying to keep sorted out which delayed reinforcer goes with which response. The cognitive load by itself precludes many such long-range associations for most species, except perhaps for special cases like food aversions for which long-range associations are vital.

But the hyperbolic function opens up another option, if circumstances permit. At sufficient delays, the later alternative is more valuable, albeit at longer delays both curves are low. A precommitting response at such times is reinforced because it precludes preference reversal. But the reinforcement is relatively slight, moving from one response at low strength to another, only marginally stronger. We may suppose that the conditioning out in the tails of the crossing curves is slow compared to that at the heights of the curves, just before pay-off. What is needed for self-control is sufficient stability in the environment to condition a marginally stronger response. A stable environment, then, would tend to overcome the bias toward impulsiveness, and quite adaptively so since impulsiveness is probably the safe bet in unstable environments. Only stable environments are likely to tip the balance towards forbearing behaviors, with their relatively slight reinforcers. Self-control is also promoted by anything that favors slower time-discounting, such as clear stimuli associating responses with delayed consequences or a self-generated system of symbols to do so. In fact, one of the self-evident evolutionary pressures for language is the benefit it provides in correlating responses with delayed consequences. The resulting reduction in I enables us to hold off temptation better, at least sometimes.

Self-control versus impulsiveness is sometimes cast in terms of rationality versus irrationality. At first look, it may seem irrational to choose a smaller reinforcer just because the larger one is later. The deeper insight of a hyperbolic theory uncovers a differing view, however. Being later, the larger one is probably less certain. It may not be delivered as promised or we may not be there to collect. Moreover, being later, it requires cognitive bookkeeping to keep track of, or to discover at all, and cognitive capacities are finite. The advantage in reinforcement for the later alternative may be more than offset by the cost in psychological effort and preoccupation. In an animal's natural economy, it may or may not be more adaptive to take the quick and small pay-off and move on. The response-strength theory spells out how the guiding dimensions interact to push the organism towards impulsiveness or self-control. The dichotomy between rationality and irrationality is replaced by continuous functions that tie behavior into current conditions and past experience. The functions themselves converge on the matching relationship for reinforced responding, adding only a time-discounting parameter to what has already been established.

7 | On the Functions Relating Delay, Reinforcer Value, and Behavior

JAMES E. MAZUR AND RICHARD J. HERRNSTEIN

Logue's (1988) thoughtful and balanced survey of different theoretical approaches to the problem of self-control omits a few conceptual and empirical distinctions that seem worthy of consideration. Our comments will focus on (1) the distinction between reinforcer value and behavioral manifestations of that value, (2) an equation that appears to provide a better description of the relation between reinforcer delay and value than Logue's equation,[1] and (3) the ability of molar maximization models to deal with the effects of reinforcer delay.

Like others who have addressed the question of self-control from a reinforcement-theory perspective, Logue focuses on the relationship between reinforcement delay and value. Her Figure 1, like Ainslie's comparable graphs (e.g., Ainslie 1975), plots value as a reciprocal function of delay. Her discussion then moves directly to implications for behavior, without any explicit statement or consideration of the relationship between behavior and value. It seems to us that there are two relationships at issue here and that it is a good idea to be quite explicit about

Originally published in *Behavioral and Brain Sciences*, 1988, 11:4, 690–691. Reprinted with the permission of Cambridge University Press. Preparation of this essay was supported by Grant MH-38357 from the National Institute of Mental Health. This chapter was a *Behavioral and Brain Sciences* commentary on a target article by A. W. Logue. It is included here because it represents a simplification and clarification of the model presented in Chapter 6.

1.

$$\frac{B_1}{B_2} = k \left(\frac{A_1}{A_2} \right)^{s_A} \left(\frac{D_2}{D_1} \right)^{s_D}$$

The exponents s_A and s_D represent the organism's "sensitivities" to amount and delay of reinforcement. The ratio s_A/s_D is a measure of degree of self-control. [Editors' Note.]

both of them. First is the one relating value (V) to delay (D), but second is the one relating behavior (B) to value. In functional form, the two relationships can be written as:

$$V = f(D, a, b, \ldots);$$

$$B = g(V, l, m, \ldots).$$

The lower-case letters in parentheses refer to variables other than delay and value which affect value and behavior, respectively, and will receive no further attention here. Among the three upper-case letters, the only directly observable dependent variable here is behavior. We know a fair amount about the nature of its functional relation to value (or its synonym, reinforcement), and that relation is the proper basis for making inferences about the unobservable variable, value.

Specifically, much evidence suggests that behavior is, at least approximately, a hyperbolic function of reinforcement, in accordance with the absolute response rate version of the matching law, as follows:

$$B_1 = \frac{kV_1}{V_1 + V_e}. \tag{7.1}$$

In this form, the parameter k represents the total amount of behavior expressed in units of B_1 and the parameter V_e is total reinforcement (i.e., value) other than that expressed by V_1. In selecting a functional form for the relationship between value and delay, it seems prudent that it be both formally consistent with the matching law, and confirmed by the available evidence. Or, to put it another way, a failure to take this logical connection into account may be a failure to take advantage of the potential for creating a coherent account of behavior.

These considerations evidently did not much influence Logue's formulation. Her equation (see note 1) stating her power-law expression for the function relating amount and delay of reinforcement to behavior, seems to us to have theoretical and empirical problems. Theoretically, it implies that the behavior ratio is as much influenced by a pair of delays of, say .1 and .01 seconds as it would be by 10 and 100 seconds, an assertion that flies in the face of plausibility. Logue recognizes the issue but does not see how readily it can be resolved by the matching law, once the right functional form for the relationship between value and delay is selected. This brings us to the empirical side of the issue.

The empirical challenge to her equation is provided by a series of experiments by Mazur (1984, 1986a, 1987), who used pigeons as subjects. Mazur found that the following hyperbolic equation offers an accurate description of the relation between reinforcer delay and value:

$$V_i = \frac{A_i}{1 + K D_i}, \tag{7.2}$$

where V_i is the value of a reinforcer delayed D_i seconds, and K is a free parameter that can vary across species, individuals, and situations (just as s_D and s_A can vary in Logue's equation). A_i reflects the amount of reinforcement, but unlike Logue, Mazur assumed only that A_i is monotonically related to physical measures of amount (e.g., milligrams of food). For his purposes, there was no reason to assume that A_i (and therefore V_i) doubles if the quantity of food or any other reinforcer doubles.

Equation 7.2 was tested by giving pigeons hundreds of choices between two delayed reinforcers, one large and one small. Mazur's procedure was designed to obtain estimates of indifference points—pairs of delay-amount combinations that had equal value. For example, Mazur (1987) found that for a typical pigeon, 2 sec of food delivered after a 6-sec delay was about equally preferred to 6 sec of food delivered after a 17-sec delay. By keeping the 2-sec and 6-sec amounts constant but varying the delays, Mazur could determine, for each increment in the small-reinforcer delay, how much the large-reinforcer delay had to be increased to maintain indifference. For all subjects, plots of large-reinforcer delays as a function of equally preferred small-reinforcer delays yielded linear functions with slopes greater than one and y-intercepts greater than zero. These functions are predicted by our Equation 7.2, but they are incompatible with other possible relations between delay and value, including an exponential relation ($V_i = A_i \cdot \exp[-K D_i]$) and an inverse relation ($V_i = A_i / K D_i$).

These empirical indifference functions are also incompatible with Logue's equation, which predicts that as the delay for a small reinforcer approaches zero, the delay for the large reinforcer must also approach zero to maintain an indifference point. However, the positive y-intercepts obtained by Mazur (1987) meant that subjects were indifferent between 2 sec of food delivered with no delay and 6 sec of food delayed a few seconds. One might argue that the delay for the 2 sec reinforcer was not really zero because it took the pigeons a fraction of a second to reach the food. However, Mazur, Stellar, and Waraczynski (1987) recently obtained similar results with rats using a reinforcer that was presumably received virtually instantaneously after a response—electrical stimulation of the brain. Even with the instantaneous delivery of a small reinforcer as one option, the rats chose to wait 4 sec or more for a larger reinforcer. This result is inconsistent with Logue's equation, and with any other equation that assumes that V_i approaches infinity as D_i approaches zero.

Let us now combine Mazur's findings concerning value as a function of delay (Equation 7.2) with the matching law in its absolute response rate form (Equation 7.1):

$$B_1 = \frac{\frac{kA_1}{1+KD_1}}{\frac{A_1}{1+KD_1} + V_e} = \frac{kA_1}{A_1 + V_e + KD_1V_e} \tag{7.3}$$

This equation differs from Herrnstein's attempt to combine the two underlying relationships (see Chapter 6). Equation 7.3 is consistent not only with the evidence on delay and matching, but also with all of the effects described in Logue's target article; it would not be appropriate here to re-review the evidence, however.

Our final comment concerns the way Logue and others have attempted to accommodate the effects of delay within the framework of molar maximization theories such as optimal foraging (Houston and McNamara, 1988). As Logue correctly states, without modification these theories predict that animals will always choose the larger but more delayed reinforcer in a self-control situation. To deal with overwhelming evidence to the contrary, the theories sometimes include time-windows beyond which delayed reinforcers are simply not counted. In our view, this approach makes little sense. For one thing, many lines of research indicate that reinforcer value declines continuously and gradually with increasing delay, not in the stepwise fashion that a time-window implies. Another way to modify a molar maximization theory would be to incorporate a temporal discounting function that lowers the impact of delayed reinforcers (cf. Rachlin et al. 1981). But whether the temporal discounting is stepwise or gradual, its inclusion in a molar maximization theory seems to be a contradiction of the meaning of the word "molar," in the absence of essentially ad hoc suppositions chosen to vitiate or disguise the contradiction. The main assumption of molar maximization theories is that some resource (e.g., food, energy) will be maximized *in the long run*, presumably because this will enhance the creature's chances of survival. Once a temporal discounting function is added to such a theory, however, the theory no longer predicts that any long-term variable will be maximized (except by coincidence). To us, the dramatic effects of delay offer clear evidence that both human and nonhuman decisions are all too frequently guided by short-term consequences, and they point to one of the major shortcomings of theories postulating long-term maximization.

8

Lost and Found: One Self

The Multiple Self, a volume edited by Jon Elster, resembles its topic. It is hard to find the core of the book, almost as hard as philosophers and social scientists find it to pinpoint the self, which difficulty is a theme of the book. In his introduction, Elster tries to pull the contributions together but does not succeed. The essay is nevertheless useful, at least to a reviewer trying to keep the pieces of an unassembled jigsaw puzzle in mind. From the standpoint of coherence or unity, the book must be judged a failure, but I enjoyed reading it anyway. It has good contributions from people with interesting things to say.

Besides Elster's introduction, there are nine essays, some but not all previously published. Likewise, some but not all were prepared originally for meetings, in 1980 and 1982, of a "Working Group on Rationality." Rationality's connection to selfhood, I hope to elucidate below. Meanwhile, there are other ways to distinguish among the essays, as follows.

Essays by David Pears, Donald Davidson, and Amélie Oksenberg Rorty deal philosophically with the ostensible paradox of self-deception and related matters, such as wishful thinking and self-contradictory, self-damaging, or otherwise irrational behavior. These essays attempt to come to grips with how a "self" can wrap itself around logical or empirical inconsistencies in belief and behavior. The question is how a person can believe, or act as if he believes, that p, yet also maintain, on logical or empirical grounds, that not p. To the extent that answers are offered, they draw mainly on the sorts of distinctions among terms that are the

Originally published as a review of Jon Elster (ed.), *The Multiple Self* (1986; references in parentheses are to this book), in *Ethics*, 98 (April 1988): 566–578. © 1988 by The University of Chicago. All rights reserved.

special concern of philosophical analysis—among beliefs, motives, and actions, among self-deception, irrationality, and weakness of the will, between logical and empirical inconsistencies, and the like.

Jon Elster also contributes an essay (in addition to his introduction) in the philosophical vein, but as literary criticism: Stendahl's purported preoccupation with the self in his fiction and in his personal life is cast as a search for a way around the paradox of wanting to be "natural" or "authentic." The incompatibility of believing (or acting as if one believes) that p and not p resembles the relationship between the desire to be natural and the hope of succeeding in one's desire, inasmuch as true naturalness implies an absence of trying to be a particular sort of person.

Two essays, by Ian Steedman and Ulrich Krause and by Serge-Christophe Kolm, attempt to absorb ostensible, systematic violations of rationality into the framework of the economic theory of utility maximization, according to which such violations are normatively prohibited. Steedman and Krause contrast the theoretical rational person guided by a single, complete, and transitive preference ordering with the multifaceted human being of ordinary experience, being pulled hither and yon by mutually incompatible tastes and value systems. If there are three or more facets with different ordinal preference structures residing within a single skin, then, as Arrow's possibility theorem proves, there is no way of averaging across the orderings of the separate facets to yield a representative aggregation. Steedman and Krause discuss how they think real people may deal with this fundamental indeterminacy, but they want to do this without deserting standard microeconomic concepts and methods. On the contrary, they hope to enrich, rather than refute, the standard economic approach to decision making.

Whereas Steedman and Krause tackle the multiplicity of "selves" (i.e., preference orderings) in a single skin, Kolm attempts to expand economic theory downward—to encompass the selfless state of Buddhist doctrine. Can utility maximization be reconciled with the belief of millions of Buddhists that the self must be denied altogether? This essay suggests that Buddhist self-denial is a way of minimizing suffering, a quite properly economic activity.

The self, in Buddhist thought as Kolm describes it, is a construct of the imagination, an illusion. Again paraphrasing Kolm, it is as arbitrary to say a self has feelings, thoughts, desires, and so forth as to say a cart has wheels, frame, axles, and so on. The cart *is* the wheels, frame, and so on; the self *is* the feelings, thoughts, and so on. Moreover, the illusion

of selfhood, which Buddhists believe arises from the gratuitous aggregation of human elements into a self, is the primary source of suffering.
Consequently, to rid oneself of much suffering, practice the discipline of
self-denial. It is only a failure of our economic imaginations, and not
a failure of economic theory, that keeps us from realizing that a self-
immolating Buddhist monk (or, for that matter, a Christian martyr) is
maximizing utility, implies Kolm.

The three remaining essays fall in an interdisciplinary category:
George Quattrone and Amos Tversky are experimental psychologists,
Thomas Schelling, an economist, and George Ainslie, a practicing psychiatrist. I would characterize their essays as naturalistic (as distinguished from philosophical or formalistic, as for the two other categories) discussions of self-deception, self-contradiction, and at least some
of the other topics already alluded to. For the final category, the point of
departure is not linguistic analysis or the extension of a formal theory,
but naturalistic observations of ordinary behavior. I would be deceiving
only myself, at most, if I claimed to be equally competent to discuss, or
interested in, the three categories here outlined. This third category is my
cup of tea, and I will therefore linger over it.

Quattrone and Tversky report the results of two experiments illustrating the human capacity for self-deception. I shall describe just one of
them. Subjects (college undergraduates) were told that their health and
life expectancies could be inferred from the effect of physical exertion on
their capacity to tolerate having a hand immersed in ice-cold water. Cold
water tolerance was measured before and after operating an exercycle.
Some subjects were told that good health correlates with a decrease in
tolerance following exercise; others, an increase. On the average, tolerance changed after exercise in the direction signifying good health. On an
anonymous questionnaire, most subjects did not acknowledge a deliberate effort to change their tolerance. They had acted as if the tolerance
change was causal, rather than diagnostic, but evidently did not know
they were doing so.

In the second experiment, too, subjects confused causality with diagnosticity. In various hypothetical elections, subjects were more likely
to vote when their voting predicted the victory of the side they favored
than when their voting carried no such predictive implications, holding
constant the actual contribution of their votes to the victory. People do
things that merely correlate with desired outcomes, when rationality requires that they only do things that contribute causally.

These examples of irrationality, Elster suggests in his introduction, are tacitly cases of self-deception: people convincing themselves that their behavior has instrumental benefits when the evidence is objectively clear that it has only correlational force. The Puritan settlers of New England fervently practiced good works even though their Calvinist belief in predestination made good works redundant. Presumably, the Puritans were confusing diagnosticity with causality, acting as if they were among the elect in the self-deceiving hope that doing so would improve their odds.

The essay by Thomas Schelling discusses "the mind as a consuming organ," which is its title. What do we consume when we watch a movie of Lassie? Schelling asks. Does "consuming" Lassie reduce our inclination to own a dog? And before you answer yes to that question, ask yourself what you consume less of after you watch *Psycho*, with its two killings. Are appetites being whetted, rather than slaked, by these goods, and, if so, what appetites? The pleasure we get from fiction is almost, but not quite, self-deception—it needs to seem real enough to suspend active disbelief but not so real as to induce true grief or terror. A roller coaster is like physical fiction—we enjoy the quasi-danger but want to be assured that the ride is safe. If the danger really seemed real, the fun would vanish.

The mind, observes Schelling, is like a computer in yet another sense than the one recently made familiar by cognitive and computer scientists. Like a home computer, the mind is a toy, in addition to being a calculating tool. Fiction, fantasy, reveries, perhaps even dreams, are examples of the mind as plaything, as consuming organ. But unlike the computer, says Schelling, "The mind is able to go from work mode to fantasy like a computer that, halfway through an income-tax programme, finds oil on your property" (Schelling, pp. 182–183).

The mind as both consuming organ and calculator is an embarrassment to the economic model of rationality, says Schelling. How can we rationally choose between eating or not eating, watching television or improving our minds with a good book, and so forth when we cannot keep our eyes off the "potato chips, sexy pictures, or animated cartoons?" (Schelling, pp. 194–195). The mind as consumer is fouling up the mind as rational decision maker (and, it could be added, vice versa). Schelling recalls the question whether creatures use genes to reproduce themselves or vice versa, and gives it a new twist: "Do I navigate my way through life with the help of my mind, or does my mind navigate its way through life by the help of me?" (Schelling, p. 195).

In other writings, Schelling had dealt more directly with the "contest for self-command" than he does here, although it remains in the near background. The ambiguity of the "self" is nowhere more vivid than in Schelling's writings. I recall, for example, something like this from one of his essays: in the days before anesthesia, patients must often have cried out to ask the surgeon to stop a painful procedure that the patient himself had requested. The surgeon probably heeded the person who asked for the operation rather than the one asking for relief from it. Who and where are these persons, why do they change places, and how does the surgeon decide who is in charge? In other situations, there is no external referee, like the surgeon, to award the contest for self-command to one self or the other, so they battle on endlessly. The person who decides, at a certain moment, to quit smoking or drinking or overeating often loses command later on to the smoker, toper, or trencherman in his skin. Resolve and indulgence may cycle indefinitely.

It is hardly surprising that George Ainslie, a psychiatrist, should concern himself with "intertemporal inconsistency," the mouthful that refers to the all-too-familiar tendency to act against one's own long-term preferences. Many addictions, obsessions, and other psychopathologies fit the mold. But it is not wholly a matter of irrationality, of acting against one's own interests, even though the examples often have that aspect as well. Consider the case of pledging contributions to one's alma mater. Fund raisers know that something less than 100 percent of pledges are made good. It is not obvious that contributing is rational and not contributing, irrational or pathological, yet the defaulter resembles the drinker or the patient. What all have in common is that short-term consequences (the pleasures of booze, the pain of parting with money, etc.) overwhelm long-term preferences—for sobriety, for supporting one's alma mater. The alcoholic may truly want to reform; the alumnus dearly loves his school. Both of them are undone by something inherent in their perspective toward deferred consequences of action.

Rational choice theorists can explain an exponential discount rate. If I lend someone something of value, or defer collecting it, then he, rather than I, can collect interest (or gain pleasure) while he holds it. Assuming a fixed interest rate per unit time, the discount rate must be exponential. Moreover, if someone else is holding a good that belongs to me, then there is a finite risk that I will not get it back. This, too, defines a fixed rate (i.e., the probability of loss) per unit time, which simply gets added to the interest rate. The theoretical result is still an exponential discount rate. An exponential discount rate does not explain intertemporal inconsistency: if you prefer chocolate cake to slimness at the moment the waiter appears with the dessert tray, you should prefer it for tomorrow's dinner as well. The pleasures of chocolate cake are less discounted than those of slimness, whether for today's dinner or tomorrow's.

Rational choice theory allows for time discounting, but not for preference reversal, for choosing the cake today and filling out a menu for tomorrow that omits chocolate cake or any other caloric dessert. The problem is that people are temporally inconsistent. They are inconsistent because they are more "myopic" than exponential time discounting permits (Strotz, 1956); they assign so weighty a role to proximal consequences that they act in violation of their own long-term preferences. It is this phenomenon, arising ultimately in the myopia of time discounting, that Schelling and Ainslie have written about so illuminatingly (see also Elster, 1984) and that accounts for the most striking challenges (in my opinion, at any rate) to the concept of an integrated self.

Ainslie's essay, "Beyond Microeconomics," enumerates a set of examples of ostensibly paradoxical human choices, some of which overlap with the paradoxes of self-contradiction and self-deception considered by the other authors. Ainslie claims to make sense of the paradoxes by elaborations of a myopic time-discounting process, which has often been observed under controlled conditions in psychological experiments. The explanations are speculative and highly original, but myopic time discounting is a fact, well established in the psychological laboratory using both human and animal subjects, and easily demonstrable by simple informal experiments. Here are some of the puzzles he addresses: (a) People not only continue to smoke after they have earnestly decided to quit but they sometimes buy cigarettes and a smoking cure at the same visit to a drugstore. (b) People abstain from congratulating themselves, from daydreaming, from masturbating, even from using a back scratcher—things they are known to be capable of doing—and choose instead to wait until the pleasures associated with those activities come from the "outside." (c) Some people lose the capacity to be reinforced by things that are ordinarily thought to be biologically endowed with the power to reinforce—food for people with anorexia nervosa, sex for frigid people, life itself for suicides.

Ainslie calls his level of analysis "picoeconomics," a step down the ladder of molecularity from the microeconomics of standard economic analysis. At the picoeconomic level, one observes the flow of individual behavior in time and finds, for example, myopic time-discounting curves. To a first approximation, the value of a deferred benefit declines in inverse proportion to its delay. (For a more up-to-date, but essentially analogous, formulation, see Mazur, 1984, 1987.) [A paragraph and figure summarizing the material of Chapter 6 have been omitted here.]

Ainslie takes his analysis well beyond the obvious implications of hyperbolic discounting. People learn, Ainslie notes, that their preferences shift. They often know that there is trouble ahead, even as they still prefer the more delayed alternative. Consequently, they feel rewarded if they take steps to preclude later reversals. Ulysses knew about myopic time discounting, hence had himself lashed to the mast (and his crew's ears plugged with wax) while the risks of approaching the song of the sirens still outweighed the temptations. Later, their "wills" would have been too weak, which is to say, they would have passed through the crossover of hyperbolic discounting curves and would have been shipwrecked on the reef near shore. Elster (1984) has written insightfully on this classical example, but Ainslie's analysis, based as it is on an imaginative extension of hyperbolic time discounting, makes the necessary contact with empirical data.

Ulysses used physical self-restraint, but other kinds of restraint are more common. The smoker or drinker or overeater, trying to stop, announces his intention to his friends. He is adding the avoidance of humiliation to the rewards for abstinence, which may or may not tip the scale toward stopping. The point of pledging is similar: to make public a long-term preference for giving, so that the sharp pain of parting with money does not destroy the resolve of the donor. The student bargains with himself, promising a night out in return for a good grade. Reneging on the internal bargain risks losing credibility with oneself, a sufficiently negative outcome to secure the bargain, at least sometimes.

People learn to ration rewards—sipping wine, delaying orgasm, and so on—in much the same way, having learned that the impulse to race to consummation is only a short-term preference for the smaller total utility. They learn to husband their drives as well as their rewards—instead of nibbling all day, when they are only marginally interested in food, they save up their hunger for a really satisfying meal later on. A connoisseur, etymologically "one who knows," knows not only *about* something but also *how* to savor it. Instead of daydreaming, we read novels and refrain from peeking at the end, which is yet another form of rationing. We wait for the compliments of others, instead of just complimenting ourselves ad libitum. All of these examples, and many others, fit the framework of [Chapter 6].

The opposing interests at different times, by-products of hyperbolic time discounting, make it possible to apply concepts from bargaining theory to the interactions within a single person. We may buy cigarettes and a smoking cure because we have made a deal with ourselves, promising one more pack before stopping. Intrapersonal bargaining often pits current appetite against a long-term alternative strong enough to sustain abstinence—such as one's credibility to oneself or a longer life expectancy. Consuming a reward, like food or sex, may be pitted against long-term interests, such as slimness or sexual purity, and lose much of its capacity to reward, as it evidently does for those afflicted with anorexia nervosa or frigidity. A person can occasionally escape from internal bargains without losing the long-term benefit by defining a non- or infrequently recurring extenuating circumstance—a rare pack of Gaulois cigarettes at the tobacco counter, for example, or a snifter of VSOP cognac on Christmas Eve.

Exponential discounting curves do not cross, unlike the hyperbolic curves [of Chapter 6]. If going to the dentist seems like a good idea at one time, it should, other things equal, seem like a good idea at other times, if time discounting is exponential. That, rationally, is how it should be, and that is why rational choice theory assumes exponential time discounting. When a person chooses inconsistently at different times, he seems, to an observer (possibly including himself) who tacitly or explicitly assumes rationality, as if he embodied multiple selves with incompatible preferences within the same skin. Some of the ostensible incoherence of the self thus arises in hyperbolic time discounting, but only because the ordinary assumption about the self (or perhaps part of its very definition) is that the self is rationally consistent.

Self as a Behaviorist Views It

In the preceding paragraph, I hinted at one point of clarification of the concept of self and its difficulties, arising from behavioral evidence—if we expect a self to be rationally consistent, then when the expectation is violated in certain ways, the notion of self is jeopardized. Here, I will broaden the point, and add other ways in which empirical evidence clarifies, perhaps even resolves, some of the difficulties noted by the authors in *The Multiple Self*.

Philosophers may wonder what empirical evidence can add to understanding, since they well know that people partition their minds in curious ways, that they deceive themselves, and that they act in violation of both their objective interests and subjective, long-term preferences. I hope to show that knowledge of concrete and specific functional relations controlling behavior qualitatively changes our view of the problem of the self. Specifically, I will present evidence and arguments supporting three propositions, the impact of which should be to explain why we intuit that there is a self but have difficulty sustaining the intuition. The three propositions are:

(1) Behavior is a field, hence integrated.

(2) Behavior is orderly but not necessarily rational.

(3) The will is not free, but it often seems to be.

Behavior is a field, hence integrated. As a physical entity, R. J. Herrnstein poses no greater or lesser problem than his car. Both of us have fuzzy boundaries. Cells and hairs slough off the former and may be replaced; tires, spark plugs, and so on come and go on the latter. One has a social security number; the other, a vehicle identification number. The former contains white cells and other living organisms that could be considered "selves" in their own right; the latter has motors, a radio, and so on that are mechanisms in their own right. The former hosts bacteria, mites, and so on that are in it or on it but not of it; the latter likewise carries passengers, bicycles, personal computers, and so forth.

To the perennial Martian observer, it would no doubt be something about Herrnstein's behavior, in contrast to his car's, that suggests personal identity for one and not the other. What is it about behavior that suggests this increment of identity? When the gasoline tank is empty, the car does not go to a filling station without a passenger in the driver's seat. A "driver" seems to be in the seat pretty much all of the time for the other one; and if that seems like too much of a claim, it may at least be said that no driver can be seen entering or leaving Herrnstein. The omnipresent driver comes close to the concept of self or will, as I understand those terms are ordinarily used.

Behavioral science has quite a lot to say about what is happening when a person gets his tank filled or otherwise behaves so as to gain some end. This is the domain of reinforcement theory, the subject most actively explored by students of operant conditioning. Behavior, speaking more loosely than we ordinarily do, is strengthened when it produces reinforcement; weakened when it produces punishment. What the punishers and reinforcers are for a given member of a given species is a complex matter that is only partially understood, but, happily, not necessary to go into here. (For a discussion of this matter, see Herrnstein, 1977a, 1977b.) For present purposes, it seems sufficient only to note that much evidence points toward a particular functional relationship between behavior and its reinforcement. (For example, see Chapter 2.)

$$B_1 = \frac{kR_1}{\sum\limits_{i=1}^{n} R_i}. \tag{8.1}$$

In the equation, B_1 represents the frequency of (or amount of time consumed by) a given class of behavior (for more precise specifications of behavioral classes, see Herrnstein, 1977a, 1977b; Skinner, 1935; Zuriff, 1985); R_1, the reinforcers produced by that behavior, measured in units of reinforcing power; R_i, all reinforcers being received by the subject during the period of observation; and k, a parameter, the dimensions of which are the same as for B_1.

Equation 8.1 is one form of what has come to be called the matching law (see Chapter 1; Herrnstein, 1970). That name reflects the implications of the equation for alternatives that differ only in the quantity of reinforcers they are producing. Variations in the quality of reinforcers or in the topography of behaviors are readily accommodated by additional parameters in the equations that change nothing conceptually (see Chapter 3; Herrnstein, 1974a). They need not concern us here. Suppose, in other words, that behavior B_1 is earning R_1 reinforcers and that behavior B_2 is earning R_2 reinforcers. Equation 8.1 then implies, other things equal, that

$$\frac{B_1}{B_1 + B_2} = \frac{R_1}{R_1 + R_2}. \tag{8.2}$$

More generally,

$$\frac{B_j}{\sum\limits_{i=1}^{n} B_i} = \frac{R_j}{\sum\limits_{i=1}^{n} R_i} \qquad (1 \cdots j \cdots n). \tag{8.3}$$

Equations 8.2 and 8.3 say that a subject, over any set of activities in the behavioral repertoire, partitions his activities in the same proportions as of the reinforcers received from those activities (assuming that activities and reinforcers have each been rendered commensurable on single scales).

The impression we have of an integrated agent of behavior in the driver's seat may, in part, arise from the character of the functional relationship between behavior and reinforcement, the equations given above or some approximation to them. These equations, and the evidence they summarize, reveal the sense in which I maintain that a subject's behavior constitutes a field. Reinforcement for any class of behavior affects the strength of all behaviors: Equation 8.1 states that the frequency of each class of behavior is directly proportional to its own reinforcement and inversely proportional to total reinforcement. Equations 8.2 and 8.3 show that the proportion of behavior allocated to a particular class is directly related to its own reinforcement and inversely related to reinforcement for other behavioral classes.

Integration across behavioral classes is made necessary because the totality of an organism's behavior in any period of time is finite, but the reinforcement is, so far as we have any reason to assume, potentially unbounded. The mapping of an unbounded variable (i.e., reinforcement) onto a bounded one (i.e., behavior) imposes formal constraints on the possible functions. For example, behavior cannot be simply proportional to reinforcement or to any positive power of reinforcement because such functions imply, impossibly, that behavior could increase without limit within a period of observation, given sufficient reinforcement.

Equation 8.1 is among the simplest permissible, given the constraints. (See Chapter 3 for a derivation.) It specifies a hyperbolic relationship between behavior and reinforcement, assuming that the reinforcement for all classes of behavior besides B_1 is held constant. Hyperbolic relationships are common in physical systems when there is the analogous constraint on the function relating unbounded input energies and bounded output activity. (For example, the so-called Michaelis-Menten equation of enzyme kinetics or the neural response of the retina to light intensity; see Mansfield, 1976.) Although the matching law was formulated as a summary of data from large numbers of experiments on animal and human subjects, in retrospect it can be seen as one, albeit not the only possible, accommodation to the objective constraint on behavior.

Behavior is orderly but not necessarily rational. The relation between reinforcement maximization and matching has been intensely scrutinized in recent years and is a matter of some controversy. But there is no disputing the observation that behavior that obeys the matching law may or may not maximize reinforcement. That is to say, if we may, for present purposes, equate reinforcement with the notion of utility as it is embodied in the theory of rational choice, then the matching law, to the extent that it is a valid principle of behavior, refutes the theory of rational choice.

This is not the proper place to rehash the arguments pro and con, but it may be helpful to describe briefly wherein matching and maximization diverge. From Equation 8.2, we may infer the following:

$$\frac{R_1}{B_1} = \frac{R_2}{B_2}. \tag{8.4}$$

The matching law states that the average reinforcement per response is equalized across alternatives. A reinforcement-maximizing law, in contrast, stipulates equalization of marginal reinforcements per response, something along the following lines:

$$\frac{\delta R_1}{\delta B_1} = \frac{\delta R_2}{\delta B_2}. \tag{8.5}$$

Although it is possible to concoct environments in which both Equations 8.4 and 8.5 hold, the equations do not entail each other. In environments in which they cannot both hold, the preponderance of evidence supports Equation 8.4, at least as I read the evidence (Chapter 10; Heyman and Herrnstein, 1986).

It is less important here whether I am reading the evidence correctly or not than that there exists a possible law of behavior that implies violations of rationality. The essays in *The Multiple Self* make it perfectly clear that behavior is sometimes irrational. This, some of them imply, casts the concept of self into some doubt. If so, it must be because the concept of self is implicitly attached to the theory of rational choice. But, for someone who accepts the matching law, irrational behavior as such ceases to be a challenge to the concept of self, only to the assumption of rationality.

The particular irrationality that Ainslie, Schelling, and others discuss as intertemporal inconsistency follows from the matching law. When the reinforcement term in Equation 8.2 is subdivided into some of its attributes, such as its rate of delivery (F), amount per delivery (A), and delay between a response and its reinforcement (D), the result may be approximated as follows:

$$\frac{B_1}{B_2} = \frac{F_1}{F_2} \cdot \frac{A_1}{A_2} \cdot \frac{D_2}{D_1}. \tag{8.6}$$

This form of the matching law is, in fact, the one used by Ainslie to derive the reciprocal relation between behavior and deferred consequences, and from which follows his account of intertemporal inconsistency.

Temporal inconsistency is not, however, the sole form of irrationality implied by matching (Chapter 10; Herrnstein, 1989). In general, it may be said that the theory of rational choice has two aspects, only one of which appears to be supported by evidence. First, it says that the allocation of behavior across alternatives tends toward states of stability in an orderly manner. Second, it says that these stability states maximize utility. The evidence, it now seems clear, supports the first principle but not the second. Those who believe, explicitly or otherwise, that the concept of self is attached to both principles are bound to run into evidence that casts the self into doubt. Writing as one who has become thoroughly accustomed to the matching law, I can say that the concept of self seems to be adequately circumscribed by stability, and by interrelatedness, as described above. In the last analysis, however, whether we feel that selfhood is or is not justified by the evidence must rest on our definitions, tacit or explicit, and I am not inclined to argue about definitions.

The will is not free, but it often seems to be.   Many of the essays in the book mention "weakness of the will" or something similar. Tying Ulysses to the mast was anticipating weakness of will. When we sneak a cigarette or a look at the potato chips, our wills have momentarily crumbled. When we deny bad news that we know is true, our wills have failed a test of strength. A Puritan doing good works in a vain effort to gain salvation is a similar case. These examples of faltering will are an essential part of the story of multiple selves that the book tells.

If weakness of will is a fitting topic for essays on selfhood, it is odd that no essay discusses free will. A few discussions of "rational agency," which seems to be synonymous, further substantiate the suggestion that rationality is tacitly assumed by some authors in their concept of self. By implication, it seems that, in addition to rationality, there is a tacit assumption of freedom of action—free will—in the concept of selfhood, at least for some of the authors. They represent well many others who might have expressed this view. No one discusses free will because it is taken for granted that the self is autonomous. The self is the driver in the machine, in most people's view, and so when people behave or otherwise express themselves inconsistently, irrationally, self-deceptively, unwittingly, or the like, it is taken as evidence that no, or no single, autonomous agent is in charge.

At least since Adam Smith, most rational choice theorists have considered their theory to be the description of the behavior of free agents.

The idea of behavioral laws for autonomous agents has always seemed self-contradictory to me. But if I am missing something, and it does after all make sense to talk about a theory accounting for the behavior of free agents, then it surely does so for the theory of rational choice and no other. The argument would presumably be that, if we were free, we would act rationally. If so, then any other law of behavior must be inconsistent with free will. For example, if the matching law is the correct theory of behavior, then it would refute the doctrine of free will because the matching law often leads to irrationality.

There seems to be quite enough evidence favoring the matching law, or something akin to it, that it becomes relevant to ask why, if behavior is not actually free, do we so often believe it is. Here are six more or less speculative reasons:

(a) The matching law often permits rational choice. For example, in environments in which the probabilities of reinforcement are independent of the allocation of behavior across alternatives (such as classical gambling situations), matching and maximization converge (see Chapter 4). Many ordinary environments are like this, or almost like this. People expect autonomous agents to maximize reinforcement, although not in those precise terms; the matching law frequently allows this expectation to be confirmed.

(b) Reinforcers are often hidden from a casual observer. Food is an obvious and visible reinforcer; the reinforcers that come from conforming to a religious teaching or a code of honor are cryptic. Behavior that is not evidently attached to an environmental consequence seems to be a purer manifestation of free will than behavior producing objective goods. We have every reason to suppose that the laws of behavior apply as well with cryptic as with obvious reinforcers.

(c) Some stimuli become reinforcers because of an individual's conditioning history—such as those associated with a religion (Wilson and Herrnstein, 1985). The qualitative power of such reinforcing stimuli is individualistic and hidden. The more individualistic a person's reinforcers are, particularly if the reinforcers are cryptic, the more free the person seems. Being subject to idiosyncratic reinforcers means being free, to that extent, of conventional reinforcers (because behavior is a field; see Brown and Herrnstein, 1975, and Wilson and Herrnstein, 1985), but it does not mean being free of control.

(d) Many reinforcers depend on mental work, which is also hidden from a casual observer, such as those involved in the examples given above of intrapersonal bargaining, and the additional ones in Ainslie's and Schelling's essays. According to ordinary criteria, one hardly ever displays one's "will power" more than one, for example, declines an invitation to play in favor of doing chores. But behavioral analysis suggests that the person may have made a bargain with himself, staking his long-term credibility to himself as a responsible person (the value of which depends on his conditioning history and possibly also on inherited temperamental characteristics) against the pleasures of the moment. The bargain making and its outcome are themselves accounted for by reinforcement, but all of the transaction is cryptic.

(e) Not only the reinforcers, but also the sources of a reinforcer, may be cryptic because they are scattered. For example, most people brush their teeth daily. If they think of the activity at all, they would think it manifested free will. After all, no one makes us brush our teeth. But if there were someone who made us brush our teeth with exactly as much compulsion as it would take to make us brush our teeth as often as we, in fact, do, we no longer would see it as free. Instead, we would see it as nearly coerced. Likewise, we may pay taxes with about as much regularity and predictability as we pay respects to the elderly. Yet, the former may seem coerced, the latter free. The difference seems to be the relative visibility of the controlling agent, not the fact of control itself.

(f) For behavior, there can be vast mismatches in the physical dimensions and scale of cause and effect. For inanimate systems, there often is a commensurability of energy for cause and effect, like one billiard ball striking another. Biological systems, in contrast, store energy, so that causes do not usually transmit energy to their effects, they transmit information. Imagine the consequences of discovering, for example, that sea air is unhealthy. Real estate values along the coasts would plunge as people migrated to the interior. Chicago would become the first, not the second, city. Because naive conceptions of causality are derived from inanimate systems, these changes in behavior would probably be attributed to free will. They would, in fact, be illustrating the power of reinforcement.

Conclusion

The ordinary concept of self, as I understand it, assumes something like unity, rationality, consistency, and autonomy. Evidence from behavioral science substantiates a limited sort of unity—the field characteristic discussed earlier—but not rationality as a rule, albeit on occasion. Empirical evidence also bears directly on consistency, which is often violated (even in the absence of any changes in "tastes") because of the myopia of time discounting. The variables controlling behavior not only are complex, they are often hidden from casual observation. Thus, behavior may seem and feel free, but not be free. Indeed, the empirical study of behavior has converged on or near a law that implies, or is at least consistent with, many of the very features of the self that the essays in *The Multiple Self* discuss so interestingly, including illustrations and violations of rationality, consistency, and unity.

When science uncovers a more complete account of an everyday concept than is contained in the common understanding, the everyday concept is either redefined or dropped. The laws of reinforcement seem to have brought us to about that point with the concept of self.

9

A Theory of Addiction

RICHARD J. HERRNSTEIN AND DRAŽEN PRELEC

<table>
<tr>
<td>We would often be sorry
if our wishes were gratified.
Aesop</td>
<td>No man ever became
extremely wicked all at once.
Juvenal</td>
</tr>
</table>

The complexity of addiction is mirrored in the many disciplines that study it. The chemistry of addictive substances falls in the domain of *biochemistry*, tolerance and withdrawal belong to *physiology*; various personality or hereditary predispositions, and the role of stressful events, are jointly addressed by *psychology* and *human genetics*; the relation to poverty, community structure, and the "social matrix" are problems of *sociology* and *political science*.

Alongside these various approaches, however, there must also be a theory of addiction that reconciles the ostensibly self-destructive consequences of addiction with the central *economic* assumption that human action can be understood as the rational pursuit of self-interest, or, if reconciliation is not possible, to examine what the implications are for that central economic assumption. The economic aspect of addiction provides the focus of this essay.

Addiction resolves into two separate, but related, paradoxes for any theory that assumes behavior to be generally utility maximizing. First, addiction is perceived to be harmful to the person who consumes the substances (above and beyond its effects on family members and society). The addict's revealed preferences are inconsistent with society's view that addiction is a losing proposition for the addict. Second, many addicts claim that they wish to change their behavior but are unable to do so. Their stated preferences are inconsistent with the preferences that they

Originally published in G. Loewenstein and J. Elster (eds.), *Choice over Time* (New York: Russell Sage Press, 1992), pp. 331–361; © 1992 Russell Sage Foundation. The authors wish to thank George Ainslie for detailed comments on an earlier draft. Thanks are also owed to the Russell Sage Foundation for support.

reveal through behavior. What shall we conclude about the relation between revealed preferences and utility in light of addiction?

Four Interpretations of Addiction

Economic theory can deal with addictive behavior in four distinct ways, as follows:

(1) *Addiction as disease, not choice.* A drug addict may be viewed as having lost the power to choose whether or not to indulge his habit, in which case the addict's behavior would not need to be accounted for by *any* theory of choice, including the economic theory. The historical shift from addictions as vices to addictions as diseases was a shift to this first approach. Similarly, criminal law has formalized a principle of "no choice" in some of its tests of culpability. The *irresistible impulse* rule, for example, allows someone to be acquitted of responsibility for an act if it can be proved (in the legal sense) that, because of overpowering emotion, the perpetrator lacked the power of choice when he committed it. Criminal law also excuses people when they act without conscious intent, or *mens rea*, on the grounds that choice requires consciousness.

Physiology has also been invoked as a reason for classifying certain behaviors as outside the domain of choice. Murderers have been defended on the grounds that their decision-making ability had been destroyed by eating too much refined sugar (the "Twinkie" defense) or by a brain tumor. The premenstrual syndrome has been offered as a defense for some crimes. Everyday theorizing about unacceptable or unconventional behavior often excuses it by calling it "physiological." The no-choice approach says, in effect, that the behavior in question was not controlled by its potential consequences.

Such arguments reflect a tendency to subtract the volitional component from choice in proportion to our knowledge of the physical reasons for the behavior. Obesity, for example, might become classified as an involuntary condition, upon discovery of its physiological correlates or genetic predispositions.

Whatever the merits of these considerations for determining personal responsibility, the "no-choice" approach does not clearly identify addictive behavior. First, behavior always has physiological reasons, whether or not we know what they are. Some relatively harmless rewards—for example, sweets—have a relatively well-understood physiology, while the rewards provided by some addictive behaviors, such as gambling, are obscure.[1] According to the no-choice approach, the former behavior would not involve choice, but the latter would, at least until we discover the physiological basis for gambling. Second, the drives for addictive behavior are not always intense, as they ought to be if the irresistible impulse criterion is to apply. Experiments have shown, for example, that obese people are more easily deterred from eating than people of normal weight by the presence of minor physical obstacles to the food. Schachter (1971), for example, describes experiments in which overweight people are more deterred from eating nuts by having to shell them or from eating sandwiches by having to get them from a refrigerator than nonoverweight people are.

Finally, addictive behavior is not distinguished by the absence of conscious deliberation. Acquiring illegal drugs, or purchasing a package tour for gambling at Las Vegas, requires more planning than many "normal" consumer choices, like watching television or hailing a taxi.

(2) *Addiction as rational self-medication.* The opposite view of addiction sees it as part of a rational lifestyle, which only appears unusual and self-destructive because we do not understand its environmental and constitutional context. A sophisticated example of this approach is provided by Becker and Murphy (1988), who treat addictive behavior as part of an optimal intertemporal consumption plan. Behavior is, in effect, *perfectly* controlled by its consequences. Addicts take full account of the impact of their current behavior on the future, including their future taste for the addictive substance, according to this theory. Although addicts may be unhappy by normal standards, they would be "even more unhappy if they were prevented from consuming the addictive goods" (Becker and Murphy, 1988, p. 691).

In all but name, Becker and Murphy depict addiction as a form of medication: The addictive commodity or activity is an expensive, inconvenient, *but nevertheless rational* treatment for special psychological conditions, such as depression, stress, and low self-esteem. This outlook seems, at least on the surface, to be inconsistent with addicts' trying to free themselves of their habit, as well as with the relative excess of young, rather than old, addicts, despite the higher long-term risks to the former.

---

1. Although a number of studies (Goleman, 1989; Roy, Adinoff, Roehrich, Lamparski, Custer, Lorenz, Barbaccia, Guidotti, Costa, and Linnoila, 1988; Roy, De Long, and Linnoila, 1989; Shaffer, Stein, Gambino, and Cummings, 1989) suggest that chronic gamblers may have low levels of activity in the noradrenergic system, abnormalities in cerebrospinal fluids, and high levels of extroversion in their personality profiles.

But, in any event, if this view is correct, then policies that restrict access to drugs and criminalize their purchase are misguided and cruel, because they penalize people whose personal welfare is already extremely low.

(3) *The primrose path.* Addictive behavior is sometimes viewed as a trap into which one is lured, because the latent costs of addiction are initially hidden or because the underlying behavioral process is deficient at making rational choices. This is an approach that splits the difference between the first two: It holds that addiction truly does depend on a person's choices, but that those choices can sometimes fail the test of rationality. Behavior *is* controlled by its consequences, according to this approach, but the result may be far from perfectly adaptive. Theories that attempt to make precise this view of addiction may be labeled *primrose path* theories. The key observation is that the typical addict goes down the primrose path believing that there is little danger of losing control (Goldstein and Kalant, 1990). The danger arises because the availability of certain substances or activities creates a situation in which a person's normal behavioral rules are inadequate. The semblance of rationality that our normal, that is, nonaddictive, behavior exhibits is then the consequence not of a utility maximization process, but rather of a good match between the behavioral rules and particular circumstances. When this match is not good—as with addictive goods—then the same behavioral rules produce poor results. We are, for example, following the same fundamental rules when we develop an appetite for golf as when the developing appetite is for heroin, according to this theory. The theory developed in the second section is a primrose path theory, but there could be others.

(4) *The divided self.* The final approach starts with the observation that the same person has different preferences at different times. In the morning, for example, a person may know that he does not want to eat dessert after supper and would order appropriately if the person could bind himself to do so; at supper, he or she succumbs. Or the person may awake daily, filled with resolve not to smoke, drink, dawdle at the water cooler, snarl at the children or spouse, etc., yet fail at virtually the first chance. It is not a matter of faulty knowledge, for the scene may be reenacted daily for years.

The self seems to be divided whenever preference depends on vantage point, which is often the case in addiction. The person discovers that too much eating or drinking has been a primrose path, leading ultimately to trouble, but this discovery fails to protect the person from succumbing when the "undesirable" alternative is at hand. Consequently, persons are in the paradoxical situation of knowing how to act in their own interest in the absence of occasions to do so, but failing to do so when the occasions arise. The person seems to know just how unwise it is to drink except when he passes a saloon.

In an article about the transmission of AIDS by unsafe drug use or sexual practices, *Science* magazine quotes Marshall Becker of the University of Michigan School of Public Health on the difficulty of changing this highly dangerous behavior: "We're asking people to make these crucial decisions over and over again at the exact moment when they're most vulnerable, which is to say when they're about to have sex or right when they're about to stick a needle in their arm" (Booth, 1988, p. 1237).

How individuals and societies do or do not resolve the paradox has inspired much of the recent work on the subject of self-control (Ainslie, 1986; Elster, 1986; Schelling, 1980; Thaler and Shefrin, 1981; Winston, 1980). The common element in formalized divided self models is that the individual is viewed as a collection of rational subagents, jockeying for control of behavior. Although individually the subagents are utility maximizers, the behavior of the collective can be severely suboptimal, as numerous examples from game theory demonstrate (i.e., the prisoner's dilemma). The third section will discuss the divided self approach to addiction, and will also show its natural links to the primrose path theory presented in the second section.

The remainder of the essay attempts to characterize addiction itself, to show its relation to primrose path and divided self theories, and, in the fourth section, to draw some implications of this analysis for public policy.

A Behavioral Model of Addiction

Four Diagnostics for Addictive Behavior

As Becker and Murphy (1988) point out, the range of activities that can at one or another time be considered addictive is extremely broad: "People get addicted not only to alcohol, cocaine, and cigarettes but also to work, eating, music, television, their standard of living, other people, religion, and many other activities" (p. 675).

Drugs, eating, television, music, a standard of living, human relationships: What do these diverse activities have in common? Some of the activities on this list would not usually be called addictions in ordinary speech, as opposed to habits, perhaps, or acquired tastes. Habits, acquired tastes, and addictions share the characteristic that they refer to activities that become more likely with repeated choice (Becker and Murphy, 1988; Leonard, 1989). But an activity is usually called an addiction only if this change seems to be a trap of some kind, locking the person into a behavior that he would avoid if he could only view it "objectively," as may well be the case with, for example, certain personal relationships and watching television.

We agree with Becker and Murphy that addictions include many activities beyond substance abuse: gambling, spending beyond one's means, compulsive buying of particular items, such as, for example, cosmetics,[2] some deviant sexual acts like exhibitionism, probably also the "type A" (hypercompetitive) personal style, excessive temper, a tendency to form self-destructive love relationships, and many others. We could perhaps include computer hacking as a new form of addiction. If it is taken as axiomatic that *all* behavior is physiologically grounded, then the everyday characterization of addictions as being "physiological" no longer makes sense. What, then, are addiction's defining features?

Although we propose here four criteria for addictions, we do not argue for a strict logical connection between the conditions listed and the use of "addiction" in ordinary language. The conditions are meant to capture most, if not all, of the denotations of the word.

(1) Addiction is normally not produced by a single action, but is rather the result of a long stream of choices.

Here, we merely note the obvious, namely, that addiction is a *habit*. A rash decision to enlist in the Foreign Legion does not constitute an addiction, no matter how long and unsatisfying the subsequent lifestyle. Addictions, in other words, are ordinarily built up from many smaller decisions; in this respect, they are closely related to what we recognize as defects in character or personality.[3] A lazy person, for example, is not

---

2. From the *New York Times Magazine* (Wells, 1988): "next to the club soda, Perrier and cat food in the refrigerator are 150 tubes of lipstick. There's hardly room for such clutter in the bathroom. That is crowded with 100 or so bottles of fragrance and uncounted cases of eyeshadow, blusher and other necessities. This shrine to makeup is Margot Rogoff's apartment. . . . This is the land of the cosmetics addict."

3. One could, in principle, become addicted from a single encounter with the addictive commodity—a super-addictive drug, for example—but even here, as should become clear, many of the characteristics of addiction, especially its harmful effects, will only materialize as the addictive behavior occurs repeatedly. An issue not dealt with in this chapter is the relation between whether or not a given addiction is thought of as betraying a defect of character and how many episodes of indulgence it takes a typical person to become an addict (but see Chapter 15).

one who *once* chose to rest rather than work, but one who is consistently predisposed to choose rest over work. We could say that he is addicted to loafing. Likewise, a hygienic person is not someone who once chose to bathe or to wear clean clothes, but someone who is predisposed to doing so repeatedly and consistently. We are likely to refer to cleanliness or idleness as addictions only if they meet at least some of the other criteria to be described, particularly the next one.

(2) Addictive behavior has significant negative intrapersonal side effects—costs that are caused by addiction but that appear in the context of other, ostensibly unrelated activities.

From the outside, it is usually clear that addictive behavior is having profound effects on the rest of the addict's life. But, for the addict, the pattern is for the psychic benefits to be directly associated with consumption of the addictive good, while the psychic costs are to be spread all around. Thus, a heavy gambler or heroin addict craves and welcomes gambling or heroin, but finds that work and family life are not as satisfying as they used to be. The negative effects of the addiction on other activities are perceived as deteriorating personal relationships or careers or the like, rather than as part and parcel of the addictive behavior—as much a part of it, say, as the pleasures of gambling or a drug "fix."

However, activities in which the primary side effect is often perceived as beneficial can also become addictive, although the addiction is then a compulsive overdoing of an otherwise worthwhile activity. With cleanliness, as with exercise or dieting, for example, the relevant benefits of greater fitness or more attractive appearance constitute a side effect, because they are absorbed as an increase in satisfaction derived from a wide range of other activities, at least initially. Beyond a certain rate of engaging in the activity, however, the side effects become negative, in that so much time is spent washing, working, or exercising that other valued activities are placed at risk. It is only at this point that the washing, exercising, or working is likely to be considered addictive.

(3) The benefits of each instance of an addictive behavior are generally more immediate than the costs.

Activities whose initial impact on the occasions when it occurs is negative (i.e., painful) are not usually regarded as addictive. If, for example, the pleasure-withdrawal cycle in opiate addiction is reversed, so that significant pain precedes pleasure, then opiates would most probably cease to be addictive. This time dependence, over rather short intervals of time, seems to play some role in many addictions. The exceptions seem mostly to be cases, such as exercise, dieting, and cleanliness, for which the natural tendency is underindulgence, perhaps because of the dominance of short-term costs over long-term benefits, but which become addictive when they shift to overindulgence, at which point they, too, may again have short-term benefits and long-term costs, albeit over longer intervals than opiates and the like.

(4) Addictive behavior, if not at first then eventually, displays temporary preference—it is anticipated with apprehension, looked back on with regret, and engaged in nevertheless.

Economic approaches to addiction invariably focus on changes in tastes—the developing of appetites for, say, alcohol, tobacco, or cocaine. While the tastes of an addict differ significantly from the tastes of nonaddicted persons, and perhaps differ, too, from his own tastes prior to repeated encounters with the addictive substance, this can also be said about anyone who regularly indulges in commodities of which the values depend on how often they are consumed. The class of such consumers is so broad that it probably includes everyone. The tastes of skiers differ from those of the general population, as do the tastes of stamp collectors, vegetarians, philanthropists, or rodeo fans. Those who love pasta have different tastes from those who love rice. The critical question is why among all these varied "acquired tastes" that constitute the preferences of any person, only a special class would be labeled addictive.

For a behavior to be called an addiction, rather than just a personal bent or appetite, it must be *unwanted*. The person must want to stop but fail to do so, or at least an onlooker suspects that the addict lacks the ability to stop should he ever want to. Most present addicts would have refused the addictive lifestyle if it had been presented to them as a one-shot choice that locks in a specific consumption program, like a regimented vacation plan.

Although our four conditions are not hard to state informally, it is difficult to characterize them within a rational choice model. This, in itself, is a significant limitation on rational modelling. To start with, the distinction between a one-shot choice and a habit does not appear in the normative theory, because a rational agent should be able to make a once and for all decision to choose a particular rate of consumption (even when the marketplace does not offer a "subscription" that locks him into that rate). Second, it should not matter to a rational individual whether the costs or benefits of an activity accrue while the activity is in progress, or whether they spread over other activities: The concept of *side effects* has no role as such in the rational model, inasmuch as the model in principle includes all effects of behavior. Third, in order to make time dependence relevant over the relatively brief cycles of pleasure and pain that occur in addiction, one would have to assume nonnegligible discount factors for intervals of days or hours or even minutes, which would then, by the logic of compound discounting, imply a complete insensitivity to consequences more than a few days or weeks away. Finally, as noted already, the idea that a person might dread engaging in certain behavior in advance, then engage in the behavior "voluntarily," only to regret having done so afterward, is on the face of it inconsistent with rationality.

Given the difficulty of expressing the four conditions within a rational model, we will turn to an alternative set of theoretical building blocks, which have been developed in the context of psychological research on human and animal choice behavior. The remainder of this section describes these concepts, and formulates a theory of addiction based on them.

Distributed Choice and Addiction

As we have just stated, standard economic theory does not draw a distinction between consumption variables that are decided with a single action (such as a car purchase), and variables that are the aggregate consequence of a series of individual, small-scale decisions, such as becoming obese or a habitual smoker. Our central hypothesis, derived from a growing body of experimental evidence, is that in this second class of situations, choice is guided by a particular sort of limited optimization, one that is fairly efficient in some situations and markedly inefficient (suboptimal) in others.

According to this idea, called the principle of *melioration* (see Chapters 4, 14; Herrnstein and Vaughan, 1980; Prelec, 1982), a person's behavior in situations of repeated choice eventually distributes itself over alternatives in the choice set so as to equalize the returns per unit invested, in time, effort, or some other constrained dimension of behavior. Behavior at the equilibrium point, where returns per unit invested are equal, conforms to the *matching law*. We have described melioration and matching elsewhere (see Part III; Prelec and Herrnstein, 1992) and will not therefore repeat it except to apply it to the question of addictive behavior. For present purposes, it should suffice to identify some commodity that is at issue—alcohol, desserts, loafing, gambling, etc.—as one of two alternatives, without saying just which one it is. The other alternative we interpret as the collection of activities and commodities that are mutually exclusive and exhaustive with respect to the commodity at issue. Thus, for drinking, one value function is for drinking, the other is simply for not drinking. The two categories of consumption are thus complementary by definition. The case for viewing alcohol consumption, and, by implication, other addictive goods, as a choice of this sort has been forcefully made by Vuchinich and Tucker (1988).

In Figure 9.1, the x-axis represents *distributed choice*, namely, the rate with which the behavior associated with the commodity or activity at issue is engaged in during some appropriate interval of time—say a week, a month, a year. If one assumes a fixed budget constraint, and as long as prices are constant, this rate is proportional to the expenditure on the item. Expenditure on everything else is proportional to the distance to the right corner of the x-axis. Changes in price have been shown to produce the expected changes in consumption rates, even for the most

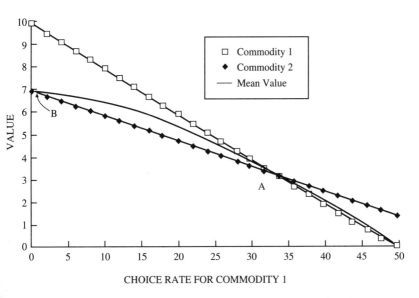

Figure 9.1   Value as a function of allocation to alternative 1, for alternative 1 (squares) and alternative 2 (diamonds). The weighted average value over both alternatives is given by the solid curve. The matching point is at A; optimal allocation is at B.

powerfully addictive substances, such as cocaine, alcohol, and tobacco (Goldstein and Kalant, 1990).

The heights of the value functions at each allocation show the average returns at that allocation, taking into account *all* the relevant parameters affecting current value, such as drive level, past history (as embodied either in pathological physiological condition or learned valuations), reinforcement-delay, price, risk, etc. The lines through the squares and diamonds represent the value of the commodity in question and its complement at any particular long-run division of choices between them (i.e., with a pair of long-run consumption frequencies). As drawn in Figure 9.1 the value functions indicate that the taste for either commodity is a linearly diminishing function of how often commodity 1 is chosen. The solid line is a weighted average of the two individual value functions, with the weight given to each function equal to the *relative* frequency of consumption of the corresponding commodity.

The process of melioration, without further elaboration, implies that choice is guided by the heights of the two value functions at any given allocation. The equilibrium pair of consumption rates is at the point where the two value functions intersect, which is also the point where the values obtained from both alternatives exactly match the mean value obtained from both (at A in Figure 9.1). This *matching* point does not necessarily coincide with the optimal pair of choice rates, which, in this example, would require total abstinence from commodity 1 (at B).

The configuration in Figure 9.1 is representative of many settings that people perceive as being problematic. The return to each commodity depends on how much of the total behavioral investment it receives. Relatively more of commodity 1 (squares) means less return to both commodities. In the context of choices between 1 and 2, average returns to 2 (diamonds) rise with its increased consumption; those to 1 decrease sharply with its consumption.

The crossing value functions in Figure 9.1, at A, define a stable matching point (see Chapter 14), in that deviations around the point are self-canceling. Matching yields much less aggregate utility than the allocation at the optimum, that is, the maximum point, B, on the average curve, where commodity 2 would be chosen exclusively, even though commodity 1 is more desirable at this allocation.

We assume that a person does not "see" the overall picture, as it is plotted in Figure 9.1, because he is psychologically embedded in it, at a certain point along the abscissa. A person who currently favors alternative 1 (e.g., at the matching point in Figure 9.1) cannot quickly test the values that would be enjoyed at the optimal distribution—or at any other distribution, for that matter. If one is in the habit of loafing, one cannot casually and instantaneously sample the fruits of long and hard labor. The interactions between consumption and taste are here modeled as a "black box," gradually shifting the value of the commodities as patterns of allocation change, and driving distributed choice toward a matching point.

It may help to think of Figure 9.1 as a representation of how a person evaluates, in reinforcement or utility terms, the two mutually exclusive and exhaustive answers, namely, "yes" and "no," to the question of consuming commodity 1 at each allocation. The question is, in principle, continuously on the table during the time interval under observation.[4]

The rate of saying yes or no intersects various of the factors affecting value. There is the obvious interaction with drive state—the average value of a slice of chocolate cake varies inversely with the rate of consumption over some range of rates. For many commodities, external availability mimics the motivational interaction. The more one visits a given berry bush, the poorer and scarcer the berries one finds.

An important class of interactions between the utility of returns and consumption rate is mediated by delay of gratification. If we say yes to cake only when it is set before us, the pleasures of eating it are likely to be forthcoming sooner than if we say yes more often, including when we are far from food, thereby initiating a chain of activities with cake at its end. It is not that we cannot initiate at any time a chain of activities with cake at the end, but that the reinforcement for doing so may be long deferred at the moment of initiation. If deferred returns are discounted, then, other things being equal, high rates of saying yes produce lower average benefits, independent of any interaction with drive or availability.[5]

How Addictions, and Addicts, Differ

Pictures like the one in Figure 9.1 may be fairly charged with hiding more than they disclose. Their advantage is that they provide a framework for taking into account the variables that affect distributed choice for

---

4. Later we show that certain cultural practices may be construed as ways of taking the question off the table at certain times, and thereby constraining the rate of "yesses."

5. An equivalent formulation of this point, more obviously linked to the principle of melioration, is that a given reinforcer spread over a longer time constitutes a lower rate of reinforcement, inasmuch as reinforcement rate is given by the ratio of the absolute value of the reinforcement to the time allocated to obtaining it (Chapter 4; Herrnstein and Vaughan, 1980).

meliorators, particularly for those commodities and activities that are generally agreed to be self-destructive. We can sample only sketchily from the vast clinical literature describing the phenomenology of addictive substances, such as smoking (tobacco), drinking (e.g., caffeine, alcohol), and the many varieties of chemical abuse (psychostimulants like cocaine or amphetamine, opiates like morphine or heroin; see Irwin, 1990, for a useful classification of addictive substances), let alone the even larger literature on all sorts of repetitive, harmful behavior. The following points reflect certain general patterns that are evident in this large literature.

In what follows, it may seem that we are simply postulating functions to fit the literature, but the point is that doing so is a natural extension of melioration, rather than a contrived use of it. This is because, unlike rational choice theory, the melioration framework makes the value functions for the competing behavioral alternatives decisive in controlling choice. In a maximization framework, a person's behavior is taken as a whole.

Individual variation.   Individuals vary in their susceptibility for falling into self-destructively high or low rates of indulgence, which is to say they vary in their propensity toward a stable matching point at which the aggregate returns are significantly suboptimal. In terms of Figure 9.1, the placement and slopes of the two value functions vary from person to person. For alcohol, the indications point toward genetic dispositions toward excessive use. But individual differences need not be a matter of narrow, substance-specific susceptibilities, genetic or otherwise. Much of the clinical literature suggests that some people are launched on the (primrose) path toward addiction or other self-destructive behavior in the spirit of problem solving (e.g., Jellinek, 1960; Orford, 1985; Pattison, Sobell, and Sobell, 1977).

A person may use a psychostimulant to deal with the problem of depression (or, in some cases, obesity). Someone else may solve a shyness problem with alcohol. A youngster may find that he or she gains the admiration of peers by smoking tobacco or dope. A gambler may have been trying to solve a genuine financial problem when he fell into the bottomless pit of compulsive gambling. In short, the motivational structures relevant to addictions and other suboptimalities can extend far beyond the specific drives involved in the activities at issue. Indeed, abusers of any given substance or activity are disproportionately likely to be abusing other substances or activities, or to be otherwise suffering from psychiatric illness (Lesieur, 1989).

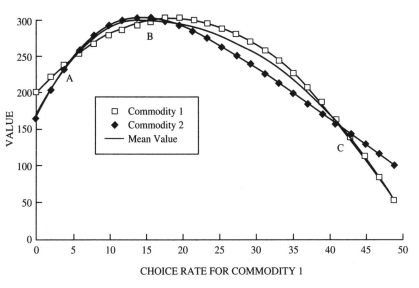

Figure 9.2   Value as a function of allocation to alternative 1, for alternative 1 (squares) and alternative 2 (diamonds). A and C are stable matching points; B is the "knife edge." The solid curve, the weighted average, virtually overlaps with alternative 2 to the left of A and with alternative 1 to the right of C.

Knife edges.   The value functions for people at risk for abusing some substance or activity may have subsidiary features that Figure 9.1 omits. It is part of the lore of certain addictions, such as that of alcoholism, that something akin to a knife edge describes the risk of succumbing (Koob and Bloom, 1988). Low levels of indulgence can go on indefinitely, but, at some threshold, the pattern tips over, and consumption seems to break free of control. Figure 9.2 is a simple alteration of the value functions consistent with this story; drinking (or some other comparable good) is represented by commodity 1 (squares), not drinking by commodity 2 (diamonds). The weighted average value is given by the solid line.

Figure 9.2 postulates two stable equilibria, at A and C, and one unstable one, B, which locates the knife edge (disregarding the two additional unstable points at exclusive choice of either alternative). The matching point at A would likely be encountered first, because it requires only a low rate of indulgence. In this region of allocation, the person is still developing a taste for drink, or still honing whatever skills may be involved in its successful use, hence, the positive slope of the value function at this level. At the same time, using alcohol may still be solving some problem in the person's life, enhancing enjoyment of activities in addition to drink itself, hence, the positive slope of the value function for alternative 2. Thus, at and around A, the average returns from drinking and nondrinking are rising with increased use of alcohol. At this matching point, drinking would seem to be under good control.

However, should something change, either in the environment or in the person's drives, so that an ongoing pattern of allocation now falls to the right of the knife edge (i.e., B in Figure 9.2), another dynamic process takes hold. Between B and C, melioration drives allocation toward dangerously high alcohol consumption, consistent with the impression of loss of control. Increased drinking begins to cast a hedonic shadow on nondrinking as the value function for commodity 2 declines. Alcohol begins to interfere with the pleasures to be had from other activities. The person may begin to experience the "anhedonia" or "dysphoria" referred to in the literature of alcoholism. Concurrently, the value function for drinking again exceeds that for not drinking, as it did at the lowest levels of use.

At some point, the returns to drink itself start declining on the average. Average returns to 1 fall off because tolerance develops to alcohol (as to most but not all addictive substances; see Koob and Bloom, 1988). The effect of tolerance is to reduce the value gained from a unit of the substance. But even without tolerance in the familiar physiological sense, it is likely that, at higher consumption rates, longer chains of activities are involved in drinking, hence, they are lower on the delay of gratification gradient (also see footnote 5). At C, which is the stable matching point at high rates of indulgence, both drinking and not drinking have lost much of their capacities to please. Neither "yes" nor "no" works well in the vicinity of C, yet there can be a stable matching point here.

Fully constrained addiction. Figure 9.2 shows how certain hypothetical configurations of functions imply equilibria that are qualitatively reminiscent of the clinical literature of alcoholism and other forms of substance abuse. In Figure 9.3, the picture is adapted to the literature of gambling. Given the disposition of the value functions here, the choice distribution would not stabilize at any point between the left or right corner, where no or all discretionary financial resources, respectively, are consumed by gambling.[6] The crossing functions, at A, are at an unsta-

---

6. To show the hypothetical value functions for a person who also has a stable matching point for low, rather than zero, levels of gambling, we need only complicate the chart slightly, borrowing the left-most end of the chart shown in Figure 9.2.

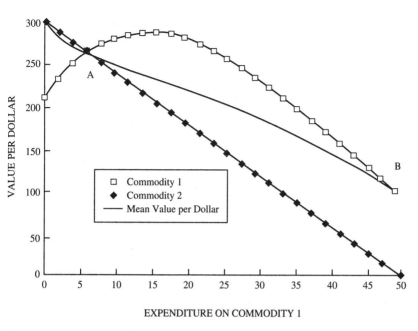

Figure 9.3   Value as a function of allocation to alternative 1, for alternative 1 (squares) and alternative 2 (diamonds). A is an unstable matching point, a knife edge, and B is the stability point for a fully constrained addiction.

ble matching point, which is to say, at a "knife edge." The right corner, B, is the point of *fully constrained addiction*, where expenditure on the addictive activity is only held in check by the sheer lack of additional resources. In hedonic terms, however, the person is still not at equilibrium.

Figure 9.3 assumes that as more financial resources are plowed into gambling (squares), the average returns from gambling will initially rise because the learning of skills or the development of tastes for the paraphernalia of gambling is rate-dependent. But at high rates of indulgence, the returns will decline if riskier bets are taken or if social disapprobation becomes part of the harvest of the activity.

Likewise, the returns from not gambling (diamonds) will also decline with increased allocation to gambling. The increasingly small resources left to the activities other than gambling are swept up by utter necessities, enough food to live rather than to enjoy, mere shelter rather than a home to take pleasure from, and so on. Little is left for other recreational activities. Instead of vacations, gifts for one's loved ones, and so on, the shrunken complement to gambling is preempted by the need to attend to often harrowing crises in one's life, many of which are themselves the by-products of gambling. Like the alcoholic or drug addict, the compulsive gambler slides toward allocations where neither gambling nor not gambling can give much pleasure.

In constructing Figure 9.3, we charged the negative by-products of gambling against activities other than gambling, rather than to gambling itself, because this is the way people tally the consequences of their actions, according to the clinical literature. Quarrels with one's spouse, triggered by gambling, are more likely to poison domestic life than to ruin gambling. In the literature of gambling, people reform precisely when they adopt a more sophisticated approach to mental bookkeeping, and place the blame where it really belongs (see Orford, 1985; Shaffer, Stein, Gambino, and Cummings, 1989).

The distinction between constrained (as in Figure 9.3) and naturally equilibrating (as in Figure 9.2) addiction is important, yet difficult to draw precisely. Financially draining addictions, such as gambling or expensive drugs, most probably leave the addict fully constrained. Food, tobacco, or video games are good examples of unconstrained addiction, for most people, at least. Alcohol consumption could be constrained or not, depending on whether the person can still draw on a steady income.

Some addictions are constrained by sheer physical restrictions on choice rate. A baseball "addict" can only watch his team play 162 different games during the regular season; a workaholic or computer addict is constrained by the clock, as are individuals caught in a "relationship addiction."

Transitions.   Other things being equal, a person who is fully constrained (Figure 9.3) should find it more difficult to break out of addiction than someone who is at an intermediate equilibrium (Figure 9.2). At or near B in Figure 9.3, not gambling would feel hedonically inferior to gambling even though virtually all resources are going into gambling. Moving leftward, toward not gambling, would require one to endure a hedonically inferior alternative for some extended period of time (until point A is reached).

The situations in Figures 9.1 or 9.2 are quite different. In Figure 9.1, a person who starts to move leftward from point A (or from point C in Figure 9.2), toward less consumption of the problematic alternative, should immediately experience an increase in satisfaction derived from *both* activities. Consequently, it may seem that only unawareness of the shape of the value functions could keep a person stuck at an intermediate, unconstrained addictive equilibrium.

But the charts have so far shown value functions only in the long run, that is to say, after each level of allocation has lasted long enough so that tastes at that point are no longer changing. Tastes interact with consumption, but the interaction may be slow, because the controlling variables for values—the basic taste changes, learning processes, social reactions to our behavior, and so on—may themselves be only slowly driven by distributed choices. For some suboptimal behavior, much, if not most, of the explanation of its tenacity probably resides in more rapid interactions with rates of consumption, the transients in the value functions as a person shifts from one allocation to another.

Goldstein and Kalant (1990) state two "special characteristics" of chemical addictants: The first is the development of tolerance, which is a decline in the hedonic yield of a given amount of the substance. Tolerance is represented in our theory as a steeply negative relation between *long-term* consumption rates and hedonic yield per unit of consumption, as shown in the earlier figures. The second characteristic is physical dependence, which leads to the *short-term* symptoms often called *withdrawal*, which we introduce at this point.

Transitory effects are especially relevant to relatively unconstrained addictions, such as smoking. The steady smoker does not experience sharp hedonic fluctuations over time, as long as cigarettes are consumed at a steady rate. The smoker, therefore, is at a point where his value functions intersect. Yet, interrupting consumption would rapidly bring on withdrawal symptoms—headaches, anxiety, tremors, and the like—which only the resumption of smoking can alleviate.

These temporary changes in value are not captured in any of the value functions considered so far. In our conceptual framework, withdrawal should appear as a shift in the value functions that is produced by a relatively brief interruption in consumption. Figure 9.4 describes a situation in which the impact of the interruption is felt as a severe reduction in the value of the nonaddictive activity. The diagram contains two value functions for this activity: The higher value function for alternative 2 (crosses) represents the satisfaction derived from nonaddictive activities *if* the consumption of the addictive substance over some relatively short time period has exceeded a threshold level. If consumption does not meet this level, then withdrawal symptoms appear, and nonaddictive activities are evaluated according to the lower function (the *withdrawal* curve, i.e., diamonds labeled 2′).

Figure 9.4 assumes, therefore, that the cost of saying no is temporarily escalated by withdrawal, hence, its net returns are down just as the

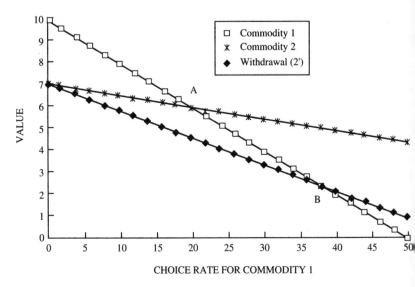

Figure 9.4   Value as a function of allocation to alternative 1, for alternative 1 (squares) and for alternative 2 if consumption of 1 is above (crosses) or below (diamonds) a threshold level. The threshold level is the level of consumption of 1 below which withdrawal symptoms appear. A is the matching point if consumption of 1 is above the threshold; B is the matching point if it is below the threshold.

person is trying to say no more often. Melioration under these conditions dictates shifting toward more of the addictive activity. The precise track of this process depends both on the time course of the short-term withdrawal symptoms and the speed with which the distributed choice process readjusts long-term tastes. In effect, the melioration process that drove the person down the primrose path to the addictive matching point is thus stacked even further against him should he try to reduce the level of indulgence.

A comparable transient influence may act on the addictive commodity itself. It has been observed that large changes in the nicotine and tar content of cigarettes may have large compensating short-term effects, but, in the long term, their effect on the consumption rates of smokers is relatively small (Gori, 1980; Russell, 1979). If smokers were regulating their habit so as to attain a preset level of nicotine or tar concentration in their bodies, as is approximately the case with certain parameters of food (e.g., calories, proteins), then it would follow that reducing nicotine (or carbon monoxide or any of the other habit-forming components of tobacco smoke) should permanently increase the number of cigarettes smoked. Instead, the long-term rate of smoking evidently sets the target levels for the habit-forming components of smoking within a broad range of consumption rates. One "gets used to" a certain physiological level of the addictant. Smoking is a self-regulating activity, but the regulatory levels are arbitrary across a range. For other "habit-forming" activities as well, a given level of indulgence sets the values of parameters around which behavior is regulated.

The opponent-process theory of motivation (Solomon and Corbit, 1974) attempts to explain this interaction between past consumption level of certain substances or activities and present motivational state by postulating homeostatic mechanisms. Substances such as nicotine, for example, induce a change in affective state. A homeostatic mechanism in the body tends to counteract that state, as part of the general tendency of the body to hold certain variables within a narrow range. If the initial affect is pleasant, the counteracting affect is unpleasant, and vice versa. The "opponent process" is affectively opposite to the original agent, but it is slower acting. As a result, after repeated exposures, it leaves the organism with a diminished initial affective reaction to the original agent, but a lingering opponent reaction.

Opponent-process theory, or something like it, handles some of the phenomena of addictions or other habit-forming commodities (Solomon, 1977; Solomon and Corbit, 1973). Rozin and his associates have suggested (Rozin, 1977; Rozin, 1982; Rozin and Fallon, 1981; Rozin and Schiller, 1980) that, for example, the taste for certain foods, such as that for chili pepper, as well as for tobacco, caffeine, and other substances, seems to conform to an opponent-process theory. The initial burning effect of chili is disagreeable, but, after one habituates to it, the lingering opponent process is pleasant, and it is the opponent process that sustains its use among habitual users. A substance that is a source of negative affect at low rates of consumption has become a source of pleasure at high rates. For substances that produce a disagreeable opponent process (known as withdrawal), such as morphine and alcohol, further indulgence in the substance itself is the antidote (as already implied by Figure 9.4 in the gap between the lines labeled 2 and 2'). Here, a high rate of consumption creates a potentially potent source of reinforcement: the avoidance of withdrawal symptoms. Whether the initial affective state is positive or negative, any attempt to decrease an established level of indulgence temporarily intensifies the desire for the substance and presumably also the reinforcement gained from consuming it.

Social and Divided Selves

The first time we let the lawn get out of hand, we may not have realized how much we will dislike the contrast between a neighbor's green grass and our green weeds. But after a few summers, we realize it all too well. Likewise, we may be warned that we are overeating, but still not know just how unpleasant it feels to catch a glimpse of our overweight selves in a mirror. Once we get there, we know. Even a meliorator, insensitive as he is to marginal returns, may know what the aggregate returns look like across a range of choice allocations.

The question is how to incorporate the effects of knowledge about the aggregate returns into the present theoretical framework. Ulysses was worldly wise enough to know the risks, ahead of time, of sailing too close to the Sirens' song and, therefore, had himself tied to the mast. Besides such individualistic solutions, the accumulated wisdom of a culture is full of warnings about the pitfalls of distributed choice. The wisdom is built into many of a culture's institutions.

Isolated and in a state of nature, we would frequently distribute choices suboptimally. The interactions between tastes and allocation are not, as a general rule, known innately, so they must be discovered by experimentation across ranges of allocations. As meliorators, however, nothing impels us to spend the time and effort looking into the black box of our taste changes. And even if we did look, we often seem to lack a natural way to keep a record of our observations. Our memories of the returns from past allocations do not usually enable us to calculate, say, just how often we should eat caviar versus hamburger in order to optimize across those alternatives. We are thus likely to ramble down the primrose path to various suboptimal equilibria, many of them far worse than overindulgence in caviar.

But in a community of meliorators, knowledge about at least some of those primrose paths would accumulate. For meliorators, it matters whether we are observing other people's behavior or just our own. Observing just our own, we are in the clutches of our own current value functions and our impoverished memories of past ones. Observing others, it would soon be noted that, as pleasing as wine may be, it can become "too much of a good thing." We would soon be saying "enough is enough," "all things in moderation," or, simply, "know thyself." Those familiar aphorisms exemplify how human culture provides meliorators like us with rules to help us cope with our suboptimalities. What follows is a sketchy attempt to classify how meliorators cope.

Social Supplementation

Society routinely exploits our sensitivity to social reinforcers. For example, we praise hard work and deplore indolence. Most people are attuned to the social reinforcers attached to their choices, and many of those reinforcers are indexed to the rates of consumption of the commodity in question. Eyebrows rise when we ask for the second, let alone third or fourth, slice of cake. Without the social supplements, many of us might work less and eat more cake than is optimal for us individually, let alone optimal for society as a whole. People who are insensitive to social reinforcers, the sociopaths, are dangerous not only to society but to themselves (Wilson and Herrnstein, 1985).

The social supplements are often keyed to circumstances—"love and marriage go together like a horse and carriage," an old song said, affirming that love (i.e., sex), like many other activities, has its proper time and place. Activities are labeled as vices, sins, evil, virtuous, etc., which is the language used to signify whether the social supplements are positive or negative and how intense they are. The concept of salvation itself is a kind of social supplement, promising benefits in the hereafter for right conduct here. When experts decide that such behaviors as gambling, excessive alcohol consumption, habitual criminality, and so on are better dealt with as "diseases" than as the "vices" they were long taken to be (Orford, 1985), an inadvertent result may be more gambling, drinking, and crime, as if a tax has been lifted from them, unless the stigma of disease is itself an incentive that fully replaces the moral incentives lost by abandoning the language of vice and virtue. By disapproving of behavior, we deter it, insofar as our disapproval has net, negative incentive value. This disregards any compensating benefits of the disease approach to bad behavior, which may be considerable.

In the framework of standard economic theory, the need for social supplements would be most readily accounted for in terms of the fundamental problem of welfare economics—that individually rational behavior may be collectively damaging. We tax the grazing of livestock on the commons and the use of other public goods, so as to avoid the "tragedy of the commons," which is to say, to deter individual freeloading. But meliorators need social supplements, too, in order to make their choices individually more nearly optimal than they would otherwise be.

As Herrnstein and Prelec (1992) show, the problem of externalities in welfare economics has an analogue in the allocations of individuals.

Because distributed choices often go haywire by stabilizing at the wrong rate, society invents many *rate-setting customs*. The optimal amount of alcohol consumption is probably not zero, but some relatively low level, at which one can enjoy some of its disinhibiting and analgesic effects, without wrecking the rest of one's life. For some people, the process of melioration fails to provide a stable matching point in this range of allocations, and the primrose path would lead to trouble, except for social practices that lend support to the desirable range of allocations. A daily cocktail or two before dinner, beer in 12-ounce bottles, a few holidays enlivened by drink, occasional religious observances with a taste of wine, are examples of rate-setting customs that detour the primrose path away from danger. A beer commercial on television advises us to "know when to say when."

There will still be some who, because of physiological constitution, past history, or present circumstance, succumb to excess indulgence even in the face of social protections against it. For alcohol, total prohibition remains a possibility, a potentially unstable matching point that has some hope of enduring when supported by the techniques of self-management that Ainslie and others have discussed. But for many bad habits, the "just say no" option is not feasible. We cannot say no to all eating, and it seems too harsh and costly to say no to all leisure, all sex, all unessential spending, and so on. Rate-setting practices, other than total prohibitions, therefore abound.

Here are some commonplace examples. For eating, the norm is three meals a day, and the meals themselves fall within a roughly prescribed range of sizes and nutritional contents. For people whose value functions might lead them to overeat, undereat, or otherwise malnourish themselves, the norm guides them toward a more desirable range of choices. Someone who eats six eggs and a cheeseburger for breakfast is likely to get immediate negative feedback from his dining companions; likewise for someone who eats no breakfast at all. For the "right" rate of leisure, we have the 9-to-5 working day, 5 days a week. The sabbath itself is an ancient and hallowed version of a rate-setting device for leisure, and also for religious observances. For charitable donations, tithing or other rules of giving help us maintain a level that we would like to choose but may find it difficult to attain because we meliorate. Daily clean clothes and a bath may stipulate a level of personal hygiene that people might otherwise frequently undershoot, to their own regret as well as ours.

In these, and many other, cases, the mechanism of control is the application of social incentives, pro or con, to constrain behavioral allocation beneficially, at least on average. But insofar as individuals vary, in value functions and in environmental circumstances, the social supplements are likely to be only crudely helpful in particular cases. Methods of self-management more finely tuned to the individual's own problems, therefore, need to be considered.

Buying a magazine or concert subscription is a way to transform one's own distributed choice into something closer to a once-and-for-all choice, once we realize that the distributed choices are suboptimal as we reckon our own utility. In effect, we buy a certain rate of allocation. Subscriptions often discount unit cost, but they probably need not do so in some instances. If we think we *should* go to the symphony or to the exercise clinic more often than we find we do, we may be willing to pay a premium for a package deal, as a way of buying an incentive to control our future behavior.

Divided Self

Once we buy an incentive to control our future behavior, we have begun to act as if more than one person, pulling in different directions, lives in our skin. Person A within us spends money for a subscription or some other rate-constraining device, hoping to influence the other person, B, also within us, because he knows from past experience that B may otherwise act against A's wishes. At different times, our preferences can be so inconsistent that we may entertain notions of demonic possession, loss of will, multiple personality, and the like. Modern theorists, particularly Ainslie, Elster, Schelling, Strotz, Thaler, and Winston, have written illuminatingly about intertemporal inconsistency. The theories explaining the inconsistency vary, but the fact itself is too much a part of everyone's experience to be in serious doubt.

Only saints and liars can say they have not experienced such problems as the following: Early in the semester, we resolve to keep up with the homework assignments, but then, having failed to do so, we run out of time to catch up and fail to earn the grade we think we could have earned. It may be a surprise to discover our failure the first time, but some students go through this disappointment semester after semester; likewise resolutions to exercise more, eat less cake, save more money, read better books, watch less television, drink less beer, and so on. These are all problems of distributed choice, involving rates of consumption that violate our own assessments of utility and rationality. They are also problems that we may, in time, discover about ourselves, and we often work hard at trying to solve them. Solutions may become extreme. For example, a newspaper story (Pitt, 1988) described a man who tried to cure his drug habit by getting himself arrested and locked up in jail, where he would be treated for drug addiction while being held in a relatively drug-free environment. He broke windows in two police stations, displayed hypodermic needles and empty crack vials in another, but found "himself back on the streets each time, his pleas ignored." A judge finally helped out by persuading a private religious organization to take him into its drug rehabilitation program.

Not only have we gone down, or at least looked down, the primrose path, but we know that we must not let ourselves be guided by the current value functions. Even after the value functions have been molded by social conventions, religious teachings, and ethical principles, we may still face trouble with our choices. It is when we know our distributed choices are suboptimal that we become conscious of a process of decision making, one that is attuned to our personal irrationalities.

Ainslie (1975) has depicted the problem of conscious decision making as one involving two hyperbolically discounted delay-of-reinforcement curves. On a given occasion, we may confront a pair of alternatives, with the larger payoff more delayed than the smaller. Because time discounting is hyperbolic (rather than exponential; see Chapter 5; Mazur, 1987; Prelec, 1989), preference may switch from the later, larger reinforcement to the earlier, smaller one, as the earlier one becomes imminent. When the alternatives are both remote, the later, larger one is again preferred.

While this theory accords with both subjective experience and experimental data about time discounting, it needs to be extended to distributed choices and primrose paths, where the problem is not so much that two alternatives have temporally displaced consequences (although they may, in many cases), but that the value functions depend on levels of allocation, as illustrated in the earlier figures, and one cannot rapidly sample from other points across the range of allocations. Someone who continually watches television 12 hours a day has a hard time discovering that life would improve with a smaller ration of watching time, because the value functions for different levels of allocation are not simultaneously available for comparison. Chapter 15 discusses some of the complexities of making choices "in the small"—for example, *a* drink versus no drink—as opposed to making them "in the large"—for example, drinking just one drink before dinner versus starting the drinking day at lunch. The former is the time discounting problem in the small; the latter is that in the large.

Addictions evidently belong to that class of behaviors for which the time discounting problem arises in the large. A person who knows he has a drinking problem, for example, is likely to want to choose a lower rate in general, but the rate he wants in the large is lower than the rate he gets by aggregating the consequences of his answers in the small. Many of the prudential rules or principles of conduct that people construct to guide their own behavior are attempts to resolve discrepancies of this form.

Conclusion: Policy Implications

It may be useful to distinguish the four approaches to addiction described in the first section by the class of policy recommendations that flow from them.

The *no-choice* approach implies a search for interventions to cure the "disease" presumably underlying the unwanted behavior. Rather than rearranging the incentives that govern choice, this approach would focus on the "root causes" of the troublesome behavior and attempt to alter them. The person should, in any case, not be held accountable for the behavior or afflicted with a sense of guilt for engaging in it, for doing so would be tantamount to trying to control behavior by incentives when incentives are, presumably, irrelevant.

The *rationality* approach implies that some people are addicts because of their endowment (where endowment includes material, constitutional, and psychological aspects). Consequently, the most efficient way to increase their welfare is to give them more money. The public policy goal for drug abusers, for example, would be to make it possible for addicts to self-medicate more easily. Disregarding the possible offsetting social welfare penalties for allowing addicts freer rein to pursue their habits, the logic of this approach argues against criminalization of addictive substances, or any other form of government action that raises prices or otherwise impedes their use (see, e.g., Becker and Murphy, 1988; Friedman, 1989; Nadelmann, 1989). It also argues against attempts to change demand by education, restrictions on the advertising of addictive substances, and the like. There should be a market for addiction insurance, that is, for insuring against the high costs of drugs, should stressful events lead to a need for drug medication. Even the restriction of substances to children violates this approach in its most thoroughgoing form. If there is an overriding social welfare argument against drug abuse even though it is individually optimal, the "rationalists" have, to our knowledge, not yet made it. The argument would take the form of showing that, even though any particular addict is making optimal choices, the utility of society as a whole is reduced by the addiction. This would not be an easy argument to make rigorously, because for every compassionately suffering loved one of an addict,[7] there may be more than one supplier of goods and paraphernalia benefiting from the addiction, along with the addict.

The *primrose path* approach suggests that society should at least provide people with more information, on the grounds that they are less likely to go down the path if they know where it is headed. A useful thought experiment in this regard is to ask whether the propensity toward addiction would be reduced by biofeedback that makes the buildup of tolerance or the imminence of physical dependence more visible. If the answer is yes, then the primrose path notion is strengthened at the expense of both of the foregoing approaches. More fanciful types of feedback might include actuarial information (vividly presented) about one's career, health, family, prospects, as a function of current behavior. Besides altering the information a person has about the long-term effects of choices, public policy could attempt to influence the value functions themselves, perhaps by applying social reinforcers to make the addictive activity less attractive in relation to its alternatives. Again, this approach splits the difference between the preceding two. It accepts the reality and significance of incentives in relation to addiction, but it does not assume that the individual can perform a rational calculation in weighing those incentives. If the first approach rejects criminalization because it is irrelevant, and the second approach rejects it because it is against the addict's self-interest, the primrose path approach may favor it for providing incentives where they could be truly helpful.

Finally, the *divided self* approach extends the analysis of the primrose path. It acknowledges that individuals can sometimes rearrange the value functions controlling their own behavior. The public policy implication is to make self-control mechanisms more accessible and more

---

7. And why are they compassionately suffering, if their addicted loved one is maximizing his or her utility?

available than they evidently are now for those who succumb. Addiction, in other words, is like market failure, in that the market does not supply convenient self-control devices (and the legal system prohibits some of them). In the absence of external precommitment mechanisms, the individual is forced to invent personal strategies for self-control (as Ainslie and Schelling, among others, have described: see Ainslie, 1975, 1986; Schelling, 1978, 1980). Public policies could be devised to make the teaching and dissemination of self-control more effective.

Part III

Against Optimization

In the following six chapters, Herrnstein strengthens the case for melioration as opposed to optimization (identified with maximization of overall reinforcement) as the mechanism of adaptation; adaptive processes discussed in this section range from biological evolution to animal choice to human decision making. It is worthwhile therefore to understand as clearly as possible where the difference between melioration and optimization lies.

A single diagram (with some variations) is repeated in the unedited version of four of the six essays in this section (what are now Chapters 10, 11, 13, and 14). It first appears as Figure 10.4b. This diagram represents a class of experimental operations (let us call this class of operations the Harvard Game) that separates melioration from optimization. In the Harvard Game, human subjects choose repeatedly between two responses (usually between pressing two buttons). The contingencies are such that optimization predicts exclusive choice of one response while melioration predicts exclusive choice of the other.

One version of the Harvard Game would work as follows: by pressing buttons, subjects earn points convertible to cash at the end of the experiment. They are faced with two buttons, A and B, and a counter indicating points earned. The experiment ends after a fixed number of button presses (say, 100), but the subject is not told when the experiment will end. The apparatus contains a ten-item register (not visible to the subject) that contains each of the subject's previous ten presses. If the subject had previously alternated between buttons the ten-item register would read: ABABABABAB. If the subject now repeats a B, then the register (dropping the first press and adding the last) would read: BABABABABB. Three rules govern the acquisition of points: (1) Each press of B earns N points; (2) Each press of A earns N+3 points; (3) N is equal to the number of B's in the previous ten presses (the number of B's in the register). These rules are not told to the subjects, who just press the buttons and observe the number of points awarded after each button press. Since the subjects are not explicitly told about the structure of the underlying reward contingencies, the subjects must infer that structure from the sequence of rewards that they observe during the experiment.

What sort of performance do melioration and optimization predict? Melioration makes the simpler prediction. Since the rate of reinforcement (points per press) is always greater (precisely three points greater) for A than B, melioration predicts that subjects will always press button A. What distribution of actions will actually earn the subject the most points? Note that although each A-press earns three more points than each B-press, each B-press enters the register and stays there for ten subsequent trials. Thus the three points gained by each A-press are more than balanced by the ten points lost on the ten subsequent trials with an A rather than a B in the register. The net loss for each A-press (except the very last ten of the session) is seven points. Thus subjects would maximize points earned by always pressing B (except for the last few presses of the session; but the subject does not know when the session will end).

What do subjects actually do under these contingencies? The answer is that they sometimes press A and sometimes press B, neither meliorating nor maximizing points earned. They shift more toward melioration or more toward maximization of points as the contingencies become more or less complex (for example, as the register becomes bigger or smaller).

As in all such tests of melioration, the Harvard Game requires that the subject choose between activities with temporally distinct types of rewards. Choice of A is followed by an immediate reward which is relatively higher than the immediate reward which follows choice of B. However, choosing B generates outcomes which are characterized by relatively higher long-run rewards. The short-run superiority of A is easy for the subjects to observe (in the example described above switching from one activity to the other always generates a three-point gap in instantaneous payoffs). In contrast, changes in long-run (or average) rate of reinforcement consequent on one choice or another are incremental, indistinct, and occur sometime in the future.

In this respect, the Harvard Game mimics everyday human self-control dilemmas. The alcoholic is in a fix similar to that of Harvard Game players. It is ostensibly better to accept a particular drink than to refuse it. If the alcoholic feels good, a drink will make him feel better; if the alcoholic feels bad, a drink will still make him feel better. But the drink goes into a "register" that accumulates not in a computer but in the alcoholic's body, making life in the future just a little bit worse.

Just as subjects in the Harvard Game do not always meliorate or always maximize points, so not all people who enjoy alcohol are alcoholics. Why not? It is instructive to examine melioration theory's explanation of why subjects playing the Harvard Game do not invariably meliorate (and why everyone who enjoys drinking does not invariably become an alcoholic) as well as an explanation by optimization theory of why those same individuals do not invariably maximize overall reinforcement rate.

According to Herrnstein in Chapter 13, "It seems that anything that makes it easier for a subject to redefine the response categories will make it easier to maximize." When choosing between A and B in the Harvard Game, subjects tend to meliorate (to choose A more than 50% of the time). But if the alternatives are restructured, not as A versus B, but more broadly as 100% A versus 90% A versus 80% A versus 70% A, and so forth, then melioration and maximization make the same prediction: 0% A is the best of these alternatives. Recall (from the general introduction) that melioration and maximization are fundamentally systems of keeping behavioral accounts. The accounting system that melioration imposes (on the theorist) adjusts response categories until a set of categories is found that explains the data. Most people who like to drink are not alcoholics (do not walk down that particular primrose path), melioration theory says, because the effective alternatives are not: take this drink versus do not take it; rather, they are: have one drink per day versus have two drinks per day versus have three drinks per day, and so forth. As Herrnstein says in the same essay, "It should be obvious that, for any situation for which a maximization strategy exists, there also exists a definition of response alternatives such that the process of melioration leads to maximization." It is in formulating such categorizations that melioration theory will have its practical use. Melioration theory can tell us how we should structure our choices so as to increase our overall well-being. In recent experiments with the procedure just described (Kudadjie-Gyanfi and Rachlin, 1996), subjects, constrained to choose in groups of three or four trials at once, chose the better long-run alternative (A, in the illustration above) significantly more frequently than when choosing on a trial-by-trial basis.

How then would an optimization theorist (for example, an economist) explain the fact that subjects in the Harvard Game do not maximize overall reinforcement rate? Candidate optimization-based explanations fit into one of two categories: explanations based on preferences, and explanations based on rational (or Bayesian) learning.

The preferences (or utility function) explanations rely on the timing of payoffs; the three points gained by choosing A in the Harvard Game are gained immediately while the ten points lost are lost later and therefore should be discounted. People who do not maximize overall reinforcement rate may yet maximize overall discounted reinforcement rate (reinforcement discounted by delay). Despite the internal consistency of this argument, most optimization theorists would immediately reject it. The short-run discount rates implied by this argument are unrealistically high. This approach requires that subjects would be willing to take three units of reward immediately in exchange for a loss of ten units of reward during the next few minutes. Some chemical addicts might be willing to make such tradeoffs, but college students (the subjects in Herrnstein's experiments) trading off cash IOU's would certainly not be willing to repeatedly accept offers like this. Short-run discount rates may be high relative to long-run discount rates (as predicted by the hyperbolic discount structure), but reasonable short-run discount rates are inconsistent with the three-for-ten tradeoff described here.

Rational (Bayesian) learning is the second category of optimization-based explanation for Herrnstein's findings. This explanation posits that decision makers hold beliefs about the world (priors) which are updated continuously and rationally as new information arrives. The updated beliefs (posteriors) are used to make sophisticated choices which are optimal *conditional on the beliefs that the decision maker possesses*. Two kinds of behavior are generated by this process: behavior which is directly reward-seeking, and exploratory behavior which is designed to elicit information which will enable the decision maker to make inferences about reward contingencies.

This framework can be used to "explain" Herrnstein's results if the theorist posits the existence of a particular prior belief. For example, consider the prior belief that button-pressing choices are unlikely to have intertemporal effects (that is, whether I pick A or B right now has no impact on the payoff contingencies that characterize future A-B choices). This prior may lead the subject to make poor choices in the experiment, but there is nothing inherently irrational about holding this prior. For example, ostensibly similar environments experienced by the subject in the past may have been characterized by intertemporal independence. Extrapolating from those past environments may be a sensible way for the subject to form priors about the contingencies in the experimental setting.

Whether melioration is consistent with Bayesian learning is still not a resolved question. If melioration behavior persists even after subjects are carefully taught the contingencies in the experimental setting or are exposed to them for extended periods of time, then melioration is inconsistent with Bayesian learning and optimization theorists will need to radically amend their current paradigm. By contrast, if melioration ceases when subjects understand the relevant contingencies, then melioration behavior can be interpreted as characterizing part of the Bayesian learning curve. In this case, optimization theorists will interpret Herrnstein's melioration experiments as valuable evidence about the structure of priors. Some experimental evidence does suggest that melioration persists even when subjects acquire information about the relevant contingencies. Specifically, "hints" seem to help performance only temporarily. Harvard Game subjects given hints (told that the immediately-worse choice might be better in the long run) meliorated slightly less at first than subjects not given hints but eventually meliorated as much as no-hint subjects (Herrnstein et al., 1993). However, this experiment has yet to be replicated and extended.

Although the relationship between melioration and Bayesian learning is an important research question, it was not the focus of Herrnstein's analysis of these experiments. The experiments were designed to address a related question regarding optimal choice: do subjects choose the best feasible actions, where best is defined with respect to the actual underlying contingencies? The chapters in this section provide clear evidence against this narrow and very strong conception of optimality. The chapters are also relevant for the understanding of broader theories of optimal choice (for example, Bayesian learning), although it is not clear whether Herrnstein's work will ultimately dethrone or refine the rational choice paradigm.

In Chapter 10, Herrnstein and Vaughan draw a parallel between intrapersonal suboptimalities and evolutionary suboptimalities (for example, genes which engender intraspecies conflict). This essay is particularly important because it is the first to present evidence on the experimental operation already described (the Harvard Game). The next four chapters extend the discussion and analysis of this experimental operation and consider some closely related and supportive experiments. Chapters 13 and 14 were written to engage economists in the discussion of these issues. Finally, Chapter 15 uses distributed choice problems and melioration as a springboard or motivation for a "psychology of legal/moral reasoning and interpretation." In this final essay, Prelec and Herrnstein argue that the behavioral suboptimalities associated with melioration provide a foundation for prudential rules which are occasionally needed to override our propensities to make flawed choices.

10 | Stability, Melioration, and Natural Selection

WILLIAM VAUGHAN, JR., AND RICHARD J. HERRNSTEIN

Optimality principles pervade the physical, biological, and social sciences (Rosen, 1967) to such a degree that it has been suggested that optimization as such is a central principle of science in general (Bordley, 1983). Within physical optics, for example, Fermat's principle of least time states, briefly, that a ray of light moving through a medium will minimize, mathematically if not dynamically, its travel time. More general is Maupertius's principle of least action: in any mechanical system in which energy is conserved, action is minimized. Still broader is the second law of thermodynamics: all closed systems achieve equilibrium by maximizing entropy. Within evolutionary biology, species are often said to be evolving toward maximization of fitness. In physiology, individual organisms are pictured as systems of adaptive homeostatic mechanisms (Cannon, 1939) shaped by natural selection to approach optimality.

Notions of optimality also have a long history in theorizing about human nature or human behavior. For Aristotle, mankind was unique for its rational soul, in addition to the vegetative and animal souls allotted to lesser forms of life. To be rational was to calculate the costs and benefits of alternative courses of action, and to behave accordingly. Jeremy Bentham, the great utilitarian, said: "Nature has placed mankind under the governance of two sovereign masters, *pain* and *pleasure*. It is

Originally published in L. Green and J. H. Kagel (eds.), *Advances in Behavioral Economics*, vol. 1 (Norwood, N.J.: Ablex, 1987), pp. 185–215. Preparation of this essay was supported by Grant nos. 586-7801 and IST-8511606 from the National Science Foundation to Harvard University. The authors are grateful for helpful criticism of an earlier draft by G. D. Homans, A. Houston, J. R. Krebs, R. D. Luce, and L. Green and J. H. Kagel.

for them alone to point out what we ought to do, as well as to determine what we shall do" (Bentham, 1789, p. 1). A rational society, he believed, would employ a "felicific calculus" to maximize pleasures and minimize pain. With the formalization of economics in the nineteenth century, rationality was more precisely translated into maximization of utility. Consumers, according to standard microeconomic theory, tend to distribute their income so as to maximize total utility. In recent times, evolutionary biology has drawn on economic analysis to promote the idea that animal behavior, as well as human, is best described within an optimality framework.

Although optimality explanations are obviously both fruitful and attractive (see Shoemaker, 1982, for a general discussion of these issues), it is our view that they are special cases of a still more general class of explanations, and that the more general class is a more appropriate heuristic for science. Systems that are said to be tending toward optimality are necessarily also tending toward stability. (We are not considering systems exhibiting limit cycles or which are chaotic, although in principle our argument could be extended to them.) The distinction between optimality and stability may, as a practical matter, be irrelevant to the physical sciences; in physics or chemistry, equilibria are not said to be better than disequilibria, only more stable. A falling stone is no longer thought to be improving itself by coming to rest on the ground, as it was in Aristotelian physics. But in the life sciences, the idea of optimization has not been merely a synonym for stability, as we will show below. As long as optimality and stability are not synonymous, it should be possible to decide whether the stability achieved by a given system is also optimal in any nonvacuous sense. Since an optimal equilibrium implies stability, but not vice versa, the normative assumption of an explanation should be no more specific than stability, with the claim for optimization requiring some further evidence beyond stability itself.

Stability analysis is largely associated with the qualitative analysis of differential equations, initiated by Poincaré in the late nineteenth century (Kline, 1972). In many cases, it is not possible to obtain analytic solutions to a differential equation. Indeed, the exact form of an equation may not be known, due to possible errors of measurement. In cases such as these, the qualitative approach is an attempt to obtain a global picture of the dynamic system under study. (Introductory discussions include Ashby, 1963, and Smale, 1980. Arnold, 1973, and Hirsh and Smale, 1974, are more advanced.)

The basic sorts of question one seeks to answer involve what states of a system constitute points of stable equilibrium, points of unstable equilibrium, and transient states. In the latter case, in what direction is the system moving? If the system is changed slightly, does the global picture change radically or not? In the study of a complex dynamic system, the theoretical framework should not unduly constrain the conclusions reached regarding the nature of the system. The assumption of optimality as used in the life sciences (particularly those dealing with behavior) seems to us unnecessarily constraining. Its thrust is often not so much examination of real-world data, but interpretation: given the data, is it possible to hypothesize intuitively a plausible dimension that is being maximized? Stability analysis, on the other hand, is more closely tied to data, as is required by the qualitative analysis of transient and stable states of a system. Rather than interpretation, the thrust of stability analysis should be experimentation (or, if that is impossible, the statistical approximation of quasi-experimentation).

The particular contrast between optimization and an alternative form of stabilization that will concern us first is in regard to the experimental analysis of behavior, where the alternatives can be illustrated by economic maximization (Rachlin, Battalio, Kagel, and Green, 1981; Rachlin, Green, Kagel, and Battalio, 1976; Staddon and Motheral, 1978), on the one hand, versus melioration (see Chapter 4; Herrnstein and Vaughan, 1980; Vaughan, 1981, 1985, 1986; Vaughan, Kardish, and Wilson, 1982) on the other. The question is: do organisms distribute their time among options so as to optimize some plausible quantity, e.g., overall rate of reinforcement, or do they tend to distribute more time to better options, even if the overall reinforcement rate suffers? According to the first view, each form of behavior must somehow be sensitive to its impact on one overriding goal, such as maximization of reinforcement rate (or value or utility).[1] According to the second view, various response alternatives literally compete with each other for the organism's time and/or energy, without regard to overall impact. The alternative theories have much in common—both of them picture a behaving organism as goal-directed or cybernetically moving toward an equilibrium state,

---

1. No theoretical distinction will be drawn here between "reinforcement," "value" or "utility," only syntactic ones that should become apparent below. It would be easy to collapse the three terms into one, but doing so might result in unfamiliar usages for the various disciplines involved.

which is to say, a stability point. Both theories belong in the tradition of reinforcement theory, with its usual implications for practical prediction and control, as well as its historical antecedents in utilitarianism and hedonism. However, since only one of them can reasonably be characterized as optimizing, the other alternative represents a sharp break with the past.

Inasmuch as maximization is a central assumption of economic theory, the confrontation with melioration must bear on economics as well as psychology. Moreover, to the extent that the formal structure of biological evolutionary selection resembles that of reinforcement, an elucidation of the abstract properties of either structure should illuminate both of them. The clash between maximization and melioration has, in fact, already been joined within biology, though not usually called by those names. (The exception is Dawkins, 1982, p. 46, who had independently proposed "meliorizing" as an alternative to biological maximization.) It should later be apparent how striking the parallel is between biology's dichotomy of group selection versus gene selection, and psychology's dichotomy of maximization versus melioration.

In contrast to the usual optimality analysis, the general analysis being suggested here does not prejudge the issue of whether or not animal behavior is optimal. The state of a dynamic system may be characterized as a point within a phase space. The geometry of that space serves to characterize both transient and stable states of the system. Presumably, some such geometry describes the transient and stable states for behavior. While various forms of stability (and instability) are possible, only certain stability points are nonvacuously optimal. The question, then, is whether the stable states are also the optimal states in any meaningful sense.

Maximization and Melioration of Individual Behavior

The assumptions of present-day economics regarding consumer behavior may be found in any microeconomics text (e.g., Mansfield, 1975). A consumer may be assumed to have a fixed income which is to be distributed among a number of alternatives, or commodities. Each commodity is such that more of that commodity, all other things being equal, increases total utility, although in a decreasing manner. In the ideal case, the consumer is assumed to have knowledge of all options and what they entail. We are ignoring questions of cardinal versus ordinal utility and risky versus certain prospects, since they are not germane to the present discussion.

The main assumption about consumer behavior is that resources are distributed so as to maximize utility. In *The Theory of Political Economy*, Jevons (1871, p. 24) stated, "I may have committed oversights in explaining its details; but I conceive that, in its main features, this theory, whether useful or useless, must be the true one." Jevons's position rested largely upon intuition and common-sense observations about human behavior (e.g., "every person will choose the greater apparent good"). More recently, intuition and common sense have been replaced by pragmatism. Samuelson (1947, p. 22), for example, says, "it is possible to derive operationally meaningful restrictive hypotheses on consumers' demand functions from the assumption that consumers behave so as to maximize an ordinal preference scale of quantities of consumption goods and services." Maximization, in short, is a useful assumption, but note that Samuelson quite properly does not suggest that it alone allows the derivation of "meaningful" hypotheses.

The analysis of consumer behavior does not usually specify the process leading to maximization of utility, but it is not hard to supply one (for more efficient single-variable optimization techniques, see, for example, Adby and Depster, 1974, or Walsh, 1975). To be concrete, suppose there are two commodities, a and b. C_a and C_b represent the total amount of income spent on a and b, while U_a and U_b represent the utility derived from each. Let $C_a + C_b = 1$ (fixed income), and the graph of U_{a+b} (total utility) as a function of C_a be unimodal with no inflection points.

Suppose a certain proportion of income is being spent on a; call this C_a^*. Then $C_b^*(= 1 - C_a^*)$ is being spent on b. This distribution of income will give rise to some total utility, U_{a+b}^*. Let C_a^* now be alternately increased by an arbitrarily small amount, to C_a^{*+}, and decreased to C_a^{*-}. The increase produces total utility U_{a+b}^{*+} and the decrease produces total utility U_{a+b}^{*-}. Now consider the following process:

$$d\frac{C_a}{C_a + C_b}/dt = f_x(U_{a+b}^{*+} - U_{a+b}^{*-}). \tag{10.1}$$

Here, f_x is strictly monotonically increasing and passes through the origin. Then, if U_{a+b}^{*+} is greater than U_{a+b}^{*-} the quantity C_a will tend to

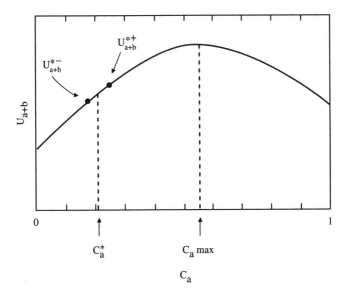

Figure 10.1   Starting at C_a^*, behavior should shift to $C_a$max for a maximizer, for at $C_a$max, total utility (U_{a+b}) is maximized. See text for explanation.

increase over time. That is, if an increase in income distributed to a gives rise to a gain in total utility, Equation 10.1 specifies that more income will be distributed to a, and vice versa. The process is shown in Figure 10.1. In this figure, overall utility, U_{a+b} is plotted as a function of proportion of income spent on a (C_a). For a particular value of C_a (labeled C_a^*), it can be seen that U_{a+b}^{*+} is greater than U_{a+b}^{*-} and hence Equation 10.1 specifies that more income should be distributed to a. The point labeled $C_a$max is such that utility is maximized; the process specified by Equation 10.1 will come into equilibrium at this point.

At equilibrium (given that exclusive preference has not been reached) we have:

$$U_{a+b}^{*+} = U_{a+b}^{*-}. \tag{10.2}$$

Since the distance from C_a^{*-} to C_a^{*+} is assumed to be arbitrarily small, Equation 10.2 is tantamount to the slope of the function for total utility being zero at $C_a$max. Given the assumptions made above regarding the shape of that function, satisfaction of Equation 10.2 is also equivalent to maximization of utility.[2]

Melioration may be characterized as follows. Again consider two alternatives, a and b, with V_a and V_b being a strictly increasing function of reinforcement per unit time obtained at each alternative. Let T_a and T_b represent times spent at a and b, with $T_a + T_b = 1$ (total time is constant). Melioration specifies:

$$d\frac{T_a}{T_a + T_b}/dt = f_x(V_a - V_b). \tag{10.5}$$

The function, f_x, is assumed to be differentiable, strictly monotonically increasing, and $f_x(0) = 0$ (see Vaughan, 1985, p. 387, for further discussion of this formulation). That is, if the value of a exceeds that of b, relatively more time will come to be distributed to a. The process specified by Equation 10.5 comes into equilibrium (given that exclusive preference has not been reached) when:

$$V_a = V_b.$$

That is, the value of a equals that of b.

For animals working on a typical operant reinforcement schedule, the assumption is made that the value of a situation is a strictly mono-

---

2. It is usually assumed that the utility gained from any commodity is a strictly monotonically increasing, negatively accelerated function of income spent on that commodity. Given this, maximization of utility is equivalent to:

$$\frac{dU_a}{dC_a} = \frac{dU_b}{dC_b}. \tag{10.3}$$

That is, C_a and C_b must be such that the slopes of the respective utility functions are equal. Suppose the slope for a were greater than that for b. Moving a small amount of money from b to a would then cause a drop in U_b that is more than offset by the gain in U_a. Only at equality is no advantageous redistribution possible.

Equation 10.3 is suggestive of the following process:

$$d\frac{C_a}{C_a + C_b}/dt = f_x\left(\frac{dU_a}{dC_a} - \frac{dU_b}{dC_b}\right), \tag{10.4}$$

where f_x has the properties specified for Equation 10.1. According to Equation 10.4, if the slope for commodity a exceeds that for commodity b, then, over time, relatively more resources will be distributed to a, and vice versa. At equilibrium, Equation 10.4 equals zero, and utility is maximized, as defined by Equation 10.3.

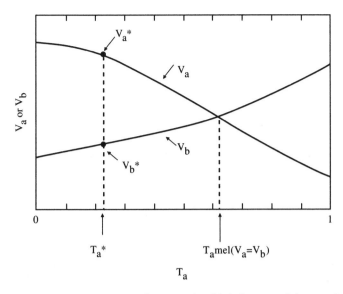

Figure 10.2 Starting at T_a^*, time allocation should shift to $T_a$mel for a meliorator, for at $T_a$mel, the reinforcing values of the competing alternatives are equalized. See text for explanation.

tonically increasing function of rate of reinforcement[3] in that situation: $V_i = g(R_i/T_i)$, where R_i represents number of reinforcements in situation i; further, $g(0) = 0$. We then have the matching law (Herrnstein, 1970):

$$\frac{T_a}{T_b} = \frac{R_a}{R_b}. \tag{10.6}$$

The operation of melioration (Equation 10.5) is shown in Figure 10.2. Here, the values of a (V_a) and b (V_b) are shown as a function of time spent at a (T_a). For a particular value of T_a (labeled T_a^*), the value of a exceeds that for b ($V_a^* > V_b^*$). Under these conditions, Equation 10.5 specifies that more time should be distributed to a. Equilibrium is reached at the point labeled $T_a$mel.

As a general rule, optimization analyses of behavior do not focus on the process (e.g., Eq. 10.1) but rather on the equilibrium state (e.g.,

---

3. The "reinforcement" of which behavior is a relatively simple and direct function may sometimes seem to be objectively specified, as by a certain number of food pellets or the like. In principle, however, it is subjective in approximately the same sense as the economic concept of "utility."

Eq. 10.2). Equation 10.2 may be deduced from Equation 10.1, given suitable boundary conditions (such as unimodality). The converse, however, is not in general true: it is not necessarily the case that the process in Equation 10.1 can be uniquely inferred from the outcome in Equation 10.2. Consider a concurrent variable-ratio, variable-ratio (conc VR VR) experiment (Herrnstein and Loveland, 1975), a procedure in which one alternative always pays off with a higher probability than the other, and the two pay-offs are otherwise identical. Although this experiment has been cited as evidence of optimization (e.g., Krebs, 1978; Rachlin et al., 1976), it is actually ambiguous with regard to discriminating between Equations 10.1 and 10.5. On conc VR VR, both an optimization process and melioration will drive an organism toward exclusive preference for the better alternative.

Figure 10.3 is a graphical representation of a conc VR VR experiment. Here R_a/T_a and R_b/T_b stand for the local rates of reinforcement in situations a and b, and $(R_a + R_b)/(T_a + T_b)$ represents overall rate of reinforcement. Here we assume that V_i is a strictly increasing function of R_i/T_i, that U_{i+j} is a strictly increasing function of $(R_i + R_j)/(T_i + T_j)$, and that T_a (proportion of time at alternative a) is equivalent to C_a (proportion of income spent on a). For any particular distribution of re-

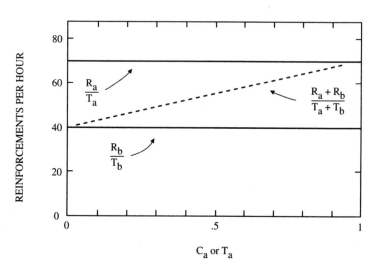

Figure 10.3 Showing a concurrent variable-ratio, variable-ratio schedule on which alternative a is more valued than alternative b. Both maximization and melioration predict exclusive preference for a.

sources between 0 and 1, V_a is greater than V_b, so T_a should increase (by Eq. 10.5). Further, an increase in C_a will increase U_{a+b} so C_a should increase (by Eq. 10.1). Under these conditions, then, both melioration and maximization predict exclusive preference for a.

Economics makes assumptions about consumer behavior not so much to study consumer behavior per se, but to facilitate deducing consequences of theories that include consumer behavior as one facet. In the large aggregations that economic theory addresses, we lose contact with individual behavior and its fine structure. Lakatos (1970) points out that scientific research programs may be characterized as consisting of a hard core, which it has been decided data cannot refute, and a protective belt of auxiliary hypotheses, which must bear the brunt of disconfirming instances. Within economics, maximization of utility by consumers is usually part of the hard core, not subject to empirical test. For behavioral analysis, on the other hand, assumptions which serve merely to protect the accepted view of consumer behavior should be dropped. The behavior of individual organisms is the ultimate subject matter for behavioral analysis, so at its hard core there should be no assumptions about the form of fundamental behavioral laws, including maximization of utility. Rather, it need merely be assumed that behavior obeys laws that can be stated at the behavioral level of analysis.

Within the study of animal behavior, data purporting to verify a maximization hypothesis have typically come from one of two areas: studies of animal behavior in the wild (in which the contingencies tend to be largely ratio-like), and studies in the laboratory employing the traditional interval and ratio schedules. The processes specified by Equations 10.1 and 10.5, however, are such that they should apply to any arbitrary schedule. (Indeed, one description of the aim of the experimental analysis of behavior is to be able, given any arbitrary schedule, to specify how the organism will behave.) Some artificially contrived schedules of reinforcement provide decisive evidence in choosing between maximization and melioration.

Figure 10.4 dramatizes how different maximization can be from melioration, given appropriate reinforcement feedback functions. In Panel A, overall rate of reinforcement (the dashed line) is constant as more or less time is spent at either alternative, so by maximization all points are points of neutral equilibrium. However, alternative a pays off better than alternative b, so, by melioration, behavior should shift toward a.

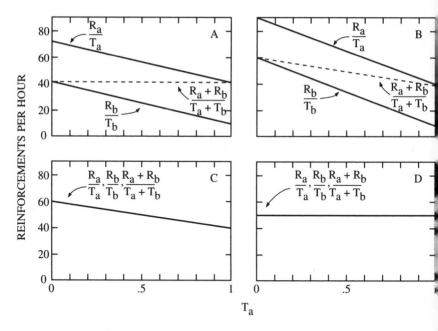

Figure 10.4   A. Reinforcement rate for alternative a (R_a/T_a) always exceeds that for alternative b (R_b/T_b). Overall reinforcement ($(R_a + R_b)/(T_a + T_b)$) (dashed line) is constant and independent of distribution of behavior (T_a). B. Reinforcement rate for a again always exceeds that for b, but overall reinforcement is maximized when b is preferred exclusively. C. Reinforcement rate is equal for a and b at all distributions of behavior, but overall reinforcement is again maximized when b is preferred exclusively. D. Reinforcement rate, for both alternatives and overall, remains constant for all distributions of behavior.

In Panel B, overall rate of reinforcement increases as behavior shifts toward b, shown by the dashed line. In addition, the same contingencies as before hold for melioration: for any distribution of time, the local rate of reinforcement (and hence the value) of alternative a exceeds that for b. If the subject is sensitive to changes in overall rates of pay, it should always tend toward b; if sensitive only to local rates of pay, it should prefer a. Maximization thus predicts exclusive preference for b, while melioration predicts exclusive preference for a.

In Panel C, the rate of reinforcement on alternative a equals that on b, so by melioration all points are points of neutral equilibrium. In addition, overall rate of reinforcement increases as behavior shifts toward b, so by maximization it should do so.

Finally, in Panel D, the rate of reinforcement is the same for both alternatives, as well as for overall rate of reinforcement. By either melioration or maximization, all distributions of behavior are points of neutral equilibrium.

Within the domain of concurrent schedules with identical reinforcers, panels A, B, and C may well exhaust the ways in which maximization may be discriminated from melioration, in the sense of having the contingencies of reinforcement for the competing responses at odds with that for responding as a whole. Although these exact experiments have not been run, the abstract principles they exemplify have all been empirically tested. Panel A corresponds to the experiment shown in Figures 5.6 and 5.7 of Herrnstein and Vaughan (1980), using concurrent alternatives with flat local reinforcement functions. Panel B corresponds to Vaughan (1981), described below (see Figure 10.8), as well as Vaughan, Kardish, and Wilson (1982), also described below. And Panel C corresponds to Figure 5.9 of Herrnstein and Vaughan (1980), in which the two local rates of reinforcement were equal to each other, but varied as a function of the distribution of behavior. It is presumably of some significance that in all of these cases, the melioration process (Equation 10.5) describes the allocation of behavior across alternatives better than the maximization process (Equation 10.1). Only in those special cases in which the two processes converge do subjects maximize (see Chapter 4).

In summary, maximization constitutes a direct translation of rationality into behavioral language. All choices are viewed as depending on their relation to one central dimension, overall utility. Equilibrium is supposedly reached when no possible redistribution of activities can increase overall utility. In contrast, melioration portrays an organism as a set of competing response tendencies, a system that is "rational" only in certain special environments (Vaughan, 1984). If one response pays off more than another, the first will increase even if the overall payoff thereby suffers. At equilibrium, all surviving responses pay off at the same average rate. Response categories that do not achieve that high a rate of pay disappear. A meliorating organism is a maximizing organism if it has an infinite capacity to redefine response categories to suit prevailing contingencies of reinforcement, for then the optimal distribution of responses in any situation would be treated as a single response category in its own right, and it would be chosen exclusively as a result of melioration (see Chapter 4). For a creature capable of learning new response configurations, melioration pushes toward maximization. However, no evidence has been provided for infinite response plasticity in any species. To the extent that the topography of response categories is not entirely determined by contingencies of reinforcement, a meliorating organism may fail to maximize.

The main virtue of stability analysis is that it focuses research on the qualitative, global properties of behavior, allowing questions regarding more precise refinements to be addressed at a later point. The major disagreement separating advocates of maximization and melioration concerns the nature of the variables governing behavior (cf. Equations 10.1 and 10.5). This is not a question requiring a precise mathematical answer. It requires a demonstration of functional relationships (or lack thereof) between environmental and behavioral variables.

For example, consider the experiment reported by Vaughan, Kardish, and Wilson (1982). Pigeons were exposed to a concurrent VI 3 min VI 3 min schedule, and distributed approximately half their time to each, an outcome consistent with both maximization and melioration. Next, a VI 1 min schedule was added, with the property that it only advanced while the pigeons worked on one of the original schedules, but delivered reinforcement (if one were available) when they were working on the other schedule. According to maximization, the pigeons should have increased their time on the side causing the third schedule to run (since this increased overall reinforcement); according to melioration, they should have spent more time on the other side (since the local rate of reinforcement on that side was now higher). The pigeons in fact shifted toward the side with the higher local rate of reinforcement, at a cost in overall reinforcement. For present purposes, the relevant feature of the experiment is that the variables purported to govern behavior by a maximization or a melioration process had opposite signs. It was not the precise outcome that mattered, but the qualitative one: more time came to be spent on one side than the other. The qualitative answer is an obvious, but essential, first step in the analysis of a dynamic system, for it precludes one of the two alternative processes.

Now consider the question of whether a slight change in a system results in an insignificant, or a significant, change in performance. On variable-interval schedules, responses are reinforced after varying and arbitrary intervals of time. Maximization holds that the moderately high response rates typically seen in pigeons derive from the positive, albeit shallow, slope of the function relating obtained reinforcement rate to response rate on the schedule, up to some rate of work at which the marginal response cost just balances the marginal gain in reinforcement. By means of a change in procedure (Vaughan, 1976), the slope of the function relating reinforcement to behavior can be made zero over a wide range of rates of work. According to maximization, the change in behavior should be dramatic; melioration would appear to predict no change in behavior (see also Prelec, 1983b). Changing the slope to zero has been shown to have little if any effect on behavior.

Natural Selection and Reinforcement

Parallels between evolution and operant conditioning have often been pointed out (e.g., Gilbert, 1972; Herrnstein, 1964; Skinner, 1981; Staddon and Simmelhag, 1971). Certain basic features of the parallel are not fortuitous, since the main question that one seeks to answer, concerning either evolution or individual behavior, is how adaptation comes about. For biological adaptation, prescientific accounts usually involved some sort of divine intervention or creation. The extraordinary adaptation of creatures to their environment has probably always been taken as evidence of God; in certain circles, it still is, as contemporary creationists exemplify. For individual behavior, prescientific accounts rely not so much on divine creation as on individual creativity or free will. Except to a small community of behavioral analysts, deterministic philosophers, and the like, adaptive behavior continues to be taken as evidence for individual autonomy.

Prescientific accounts of biological or behavioral adaptation postulate an agency outside of natural law, divine creation, or free will, beyond science and hence not subject to its causal analysis. Not surprisingly, a scientific analysis moves the agent back within the domain of natural law, back into biology or psychology. Darwin's theory of evolution noted that individuals vary along a number of dimensions. If given phenotypic values confer a greater ability to survive and reproduce than others, and are based on genetic differences, then selection will occur. Over time, this mechanistic process leads to adaptation. What God is said to do in an act of creation, evolution does, though presumably more slowly, by the sifting process of natural selection.

Variation and selection also provide a key to behavioral adaptation. A single organism's behavior varies, some of it leading to outcomes that influence the strength of those behaviors. The notions of reinforcement and punishment play the same roles in behavioral analysis as reproductive success and failure play in evolutionary biology. Selection thus again operates over time to produce adaptation by a mechanistic process; free will is said to achieve the same result instantaneously by insight.

The parallel between evolution and individual behavior goes beyond variation and selection. In each case, prescientific accounts serve both to seem to answer vexing questions and to give people an inflated sense of their worth (in the case of evolution) or their options (in the case of individual behavior). It is therefore not surprising that reactions to the scientific account tend to be virulent. John Herschel, a contemporary of Darwin, referred to natural selection as "the law of higgeldy-piggeldy" (see Hull, 1973). Koestler (1968) has referred to behaviorism as "a monumental triviality."

On the face of it, adaptation by means of variation and selection would appear to be a recipe for maximization—of fitness in the case of evolution, and of reinforcing value in the case of behavior. Nevertheless, a closer look at the process of evolution leads to a different conclusion, just as a closer look at behavior does.

Natural Selection and Stability

In an introductory chapter, Davies and Krebs (1978, p. 2) discuss two approaches to theorizing within behavioral ecology. The first is optimality theory: "natural selection should tend to produce animals which are maximally efficient at propagating their genes and therefore at doing all other activities." They then point out a limitation of optimality theory: an individual's best solution to a problem may depend on what others in a population are doing. As a result, a different approach is warranted, which is to find what is termed an evolutionarily stable strategy (ESS). This consists of a strategy such that if most members of a population adopt it, it is not susceptible to invasion by some other strategy. That is, the system is stable. Whether, in addition to being stable, an ESS is also optimal for a species is a question we reserve for later, but note here that the contrast between optimality and an ESS has essentially the same dimensions as that between utility maximization and melioration, in each case pitting a rule for optimal stability against a rule for stability per se.

The use of optimality considerations in evolution is usually traced to Wright (1932) (although McCoy, 1979, points out that Janet, 1895, made similar proposals). Wright pictured a species as occupying an area in a two-dimensional "landscape" onto which were projected contours of equal adaptive value. In general, evolution would consist in the species climbing the steepest slope in its vicinity. While picturesque, this conception of maximization has serious defects (see, for example, Maynard Smith, 1978). Given a number of adaptive peaks, where a "hill-climbing" species lands depends on initial conditions and may be a local, rather than global, maximum. The hill-climbing metaphor fails to provide a credible model of global maximization. It actually comes closer to being a model of an ESS, although, for an ESS, the genes, rather than the species as a whole, must do the "climbing."

If, in contrast, what one means by maximization is simply maximization (however achieved) of the mean reproductive fitness of individuals across a population, then one is driven to accept group selection, for a conspecific's offspring would be as potent in selection as one's own. From the standpoint of mean reproductive fitness, even one's own genes would have no special status in comparison with any conspecific's. Based on numerous observations which appeared to suggest that animals behave altruistically (for the good of the group), Wynne-Edwards (1962) argued at length for the pervasiveness of just the sort of group selection that would maximize mean reproductive fitness. But his argument ultimately failed for two reasons, as other optimalists soon pointed out. A more limited form of altruism was shown to be derivable from kin selection, which is, in turn, derivable from individual rather than group selection (Hamilton, 1964). To the extent that a kin's genes match one's own, this more limited altruism reduces to egocentricism, as far as natural selection is concerned. Moreover, a population of altruists is in any case unstable in principle, inasmuch as genes for selfishness, should they arise, will spread throughout the population (Dawkins, 1976).

Besides the problems with group selection based on mean fitness maximization, the very concept of maximizing adaptiveness is plagued with difficulty. Williams (1966), exemplifying the traditional optimalist view, says, "Natural selection would produce or maintain adaptation as a matter of definition. Whatever gene is favorably selected is better adapted than its unfavored alternatives" (p. 25). But what happens as a matter of definition tells us nothing about the world. Williams goes on to say that "the theoretically important kind of fitness is that which promotes ultimate reproductive survival" (p. 26). But can a mechanistic selective process conceivably promote "ultimate" survival? Consider, for example, Hardin's (1968) discussion of the tragedy of the commons, the risk of irreversibly overutilizing the environment. At some point during human evolution, man survived while closely related forms did not, presumably a reflection of man's greater fitness. Suppose that one component of that fitness was having sufficient intelligence to graze animals or build factories. But now, it turns out, mankind's intelligence or forbearance may be insufficient to prevent its own demise through overexploitation of the collective resources. In such a context, "ultimate reproductive survival" is a paradoxical concept at best, for a less bright species would not have grazed animals or built factories.

The above discussion has implicitly assumed that the fitness of a particular genotype is independent of the relative frequency of the genotype within a population. An analysis of the alternative, frequency-dependent fitness, was initiated by Fisher (1930), who sought an answer to why the ratio of the two sexes usually approaches one to one. A traditional optimalist might have answered that an equal number of males and females is good for the species, since it maximizes the probability of opposite sexes meeting, hence mating, or that females should greatly exceed males, if each male can fertilize many females. Fisher (making some assumptions about the transmission of gender that we may neglect here) reasoned that, if a population contained more males than females, females would then have greater fitness than males (inasmuch as the average female's genes would be more highly represented in the next generation than the average male's). The opposite argument applies if females outnumber males. Hence, any deviation from a one-to-one ratio (or possibly some other stable ratio, if parental investments are taken into account) creates conditions which automatically tend to reestablish that ratio. An equilibrium sex ratio should be approximated whether or not it is good for the species.

Since Fisher's analysis of sex ratios, the notion of frequency-dependent selection has established itself firmly. A convenient example is Maynard Smith's (Maynard Smith, 1974, 1976, 1982; Maynard Smith and Price, 1973) game-theoretic analysis of animal conflict. Why, Maynard Smith wondered, are conflicts often settled without actual bloodshed? On the face of it, fighting one's opponent to the death would appear to confer greater fitness on an individual than allowing a vanquished opponent to escape. But from the group selectionist standpoint, ritualized combat may seem advantageous inasmuch as it allows the combatants to survive and reproduce. In fact, as Maynard Smith showed, neither the simple individualistic nor the group selection answer realistically captured the actual selective contingencies.

Maynard Smith's approach is exemplified by a hypothetical species (see fuller discussion in Maynard Smith, 1982, and Dawkins, 1976), individuals of which may use either one or the other, but not both, of two strategies when confronting another individual of the same species.[4]

---

4. For our purposes, it is sufficient to consider only the case in which the strategies are mutually exclusive for any particular individual and reproduction is asexual; fuller treatments also attempt to deal with individuals that can employ either strategy and with sexual reproduction.

We assume the strategy used is under genetic control. The first strategy is termed "hawk" and is characterized by the individual always fighting until one of the combatants wins and the other loses; the loser is always injured. The second strategy is termed "dove." If a dove meets a hawk, it runs away; the hawk wins the contest, but neither is hurt. If a dove meets a dove, both posture for a while. Both doves lose time, and one wins the contest and the other loses, but neither is injured.

Next we must set values relating the outcomes of contests to fitness. Let us assume that winning a contest confers 50 fitness points on the winner and 0 on the loser; that an injury costs 100 points; and that time spent posturing costs 10 points. If a hawk meets a hawk, one will win and the other will lose and in the process be injured. The average payoff to a hawk meeting a hawk is therefore -25 ($= (+50$ for winner $- 100$ for loser)/2). If a hawk meets a dove, the hawk wins the contest, hence gains 50 points, while the dove runs away and loses 0. If a dove meets a dove, both waste time posturing; in addition, one wins and one loses. The average payoff for a dove meeting another dove is therefore 15 ($= (+50 - 20)/2$).

Now imagine a population consisting either of all doves or all hawks. In a population of doves, the average payoff to each individual is 15 points. But if a mutant hawk appears, it will win all contests. Hence, in a pure dove population, a hawk always receives 50 points, so hawk genes would tend to sweep the population. In contrast, in a population of hawks, the average payoff to each individual is -25. If a dove mutant appears, its average payoff is 0. Since this is greater than -25, dove genes will tend to sweep the population. Both pure dove and pure hawk populations are therefore unstable: a mutant with the opposite strategy will have greater fitness than the population average. Figure 10.5 shows the fitnesses for hawk and dove as a function of the proportion of the population playing dove. Given the particular parameters used here, a stable mix of hawks and doves would have 5/12 doves and 7/12 hawks.

From Figure 10.5 it can be seen that, at stability (evolutionarily stable strategy, or ESS), the average fitness for each individual is $6\frac{1}{4}$, independent of whether the individual is a hawk or a dove. The average fitness for each individual at all proportions of doves and hawks is plotted as the dashed curve. The equality of fitness for doves and hawks is a necessary (though not sufficient) condition for stability: if they were not equal, the strategy with greater fitness would subsequently constitute a higher proportion of the population, until equality was achieved. Note further that the average fitness at equality of $6\frac{1}{4}$ is suboptimal for the group; it

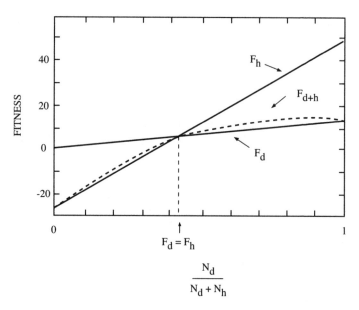

Figure 10.5   Fitness for "doves" (F_d) and for "hawks" (F_h) as a function of the proportion of doves in the population. N_d and N_h refer to the number of doves and hawks, respectively. At a fitness of $6\frac{1}{4}$, the two forms equalize in average fitness. The evolutionarily stable strategy (ESS) thus predicts 5/12 as the dove proportion. The total average fitness across doves and hawks (F_{d+h}) is given by the dashed curve.

is, for example, less than what could be achieved in an all dove population, which is 15. In this example, stability and optimality, in the sense of the maximum average fitness per individual, are mutually incompatible.

It could be countered, at this point, that, even though the species as a whole is not maximizing fitness, competing forms are in fact doing just that: given the context of individuals, the hawk or dove gene would each lose fitness by increasing its proportion above that at the ESS. But that, in fact, would be to grant that evolution is a melioration, rather than a maximization, process. The parallel has been drawn between an individual with its various responses and a species with its various genes. Just as an individual's responses do not act in concert for the benefit of the individual, so the genes of a species do not act in concert to increase the species' fitness. In both cases, equilibrium is achieved, not necessarily at an optimum, but at a point where the meliorating elements are equally benefited, which is to say, at the point specified by the matching law.

In a discussion of the competitive exclusion principle, Ayala (1972) has discussed the necessary and sufficient conditions for two competing species to achieve a stable mix, as distinguished from one species excluding the other. For a stable mix, there must, first, exist a mix such that individuals of the two species have equal fitness, which means that over time their relative numbers remain constant. Second, given a deviation from that mix, the species whose relative representation has increased must simultaneously move to a lower individual fitness than the other: a deviation from equality drives the population back toward a mix that restores the equality point for fitnesses. If either or both of the two conditions are not met, in the long run one of the species will competitively exclude the other. Similar dynamics govern the competition between genetic polymorphisms within a species.

Frequency-dependent selection has been shown empirically to lead either to stable mixes or exclusions (for reviews, see, for example, Ayala and Campbell, 1974; Clarke, 1979), illustrated by Batesian mimicry and Mullerian mimicry, respectively (Fisher, 1930). In Batesian mimicry, a palatable species resembles a noxious one, thereby gaining in fitness by being less often taken as prey. Now consider a genetically polymorphic species with two phenotypes, one exhibiting mimicry of a noxious second species, and the other, of a noxious third species. The two forms are not themselves noxious, but they gain fitness by resembling noxious species. Whenever the proportion of one of the two forms rises, its ratio to the mimicked species also rises, other things being equal. Predators should then tend to favor it over the other phenotype, because its probability of noxiousness has fallen. As predators consume more of the first phenotype, its probability of noxiousness rises, reducing consumption, and so on. The two forms thus tend toward a stable mix, the specific value of which depends on the various parameters.

In Mullerian mimicry, two noxious species resemble each other, so that a predator that has encountered one will tend to avoid both. Here, the species is itself noxious and so is the species it resembles. Consider a noxious polymorphic species showing Mullerian mimicry for one or the other of two other noxious species. If one of these forms is more frequent than the other, it will tend to increase in frequency, since predators will have more often encountered it and discovered its noxiousness (disregarding effects of variations in degrees of noxiousness). At stability, only the more frequent form remains.

Harding, Allard, and Smeltzer (1966) discuss a case of frequency-dependent selection in lima beans. Over generations, plants that were heterozygotic at a particular locus tended to constitute about 7% of the otherwise homozygotic population. Experimental tests showed that heterozygotes had lower fitness in the presence of other heterozygotes than in the presence of homozygotes. The authors suggested that the frequency-dependence might have resulted from competition between the heterozygotes. At about 7% heterozygosity, fitness for heterozygotes approximately matched that for homozygotes.

Since frequency dependence is self-evidently an essential consideration in selection, it has attracted considerable theoretical, as well as empirical attention (e.g., Slatkin, 1978; Ayala, 1972; Dawkins, 1976, 1980, 1982; Clarke, 1979). For our purposes, however, the detailed complications are beside the point. Instead, we shall represent Ayala's necessary and sufficient conditions for stable mixes graphically (see Figure 10.6), by plotting the fitness differences between individuals in two competing species (or two competing forms in one species) as a function of the relative sizes of the two populations, N_a and N_b. In order to achieve a stable mix, the interaction between the species or forms must conform to a function that intersects zero on the ordinate in Figure 10.6 with a negative slope (or a discontinuous approximation to a negative slope) at the intersection. The actual function drawn by the solid line is, in fact, derived from the hypothetical hawks and doves (see Figure 10.5). It has already been shown that, in this example, stability is suboptimal.

As long as fitness a exceeds fitness b, the population will shift toward a, and vice versa. If the fitness-difference function intersects zero with positive slope, then fitness-equality will be unstable, since deviations on either side of equality are self-amplifying (see lower dashed function in Figure 10.6). Under these conditions, the population will become totally a or b, depending on initial conditions. Besides intersections with positive or negative slope, the fitness-difference function may simply not intersect zero, illustrated by the upper dashed function in Figure 10.6. Here, the a's would prevail, and it would look like a case of frequency-independent selection, although, strictly speaking, the function should at least have zero slope to justify that term (see discussion of Figure 10.7, below).

In a case of strict frequency-independent selection, the stable state is also necessarily optimal in the maximization-of-fitness sense. The more fit form simply excludes the less fit form. However, for the other kinds of stability, whether with an ESS mix of the two forms or competitive exclusion by one form, maximization of fitness depends on aspects of the fitness-difference functions that have no necessary bearing on the stability point. Figure 10.7, for example, contains three sets of contingencies that would yield populations of all a's given the principles of competi-

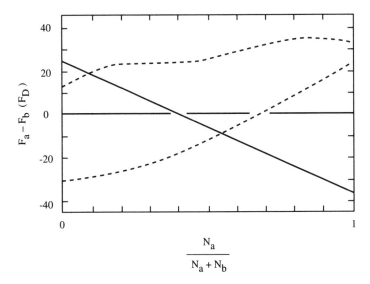

Figure 10.6   Difference between fitness of form *a* and form *b* (F_D) as a function of proportion of the population adopting form *a*. Only the solid line (sloping downward to the right) illustrates the condition for a stable mix of two populations. That line plots the fitness-difference function for hawks and doves (see Figure 10.5). Lower dashed line is unstable because any deviation from a fitness difference of zero is self-amplifying. Upper dashed line predicts exclusion of *b* by *a*.

tive exclusion. In each panel, form *a* has greater individual fitness (F_a) than form *b* (F_b) at all relative frequencies, so the population would shift to *a*'s exclusively. However, only for panel B does an all *a* population maximize overall fitness (F_{a+b}); for the other panels, an all *a* population minimizes overall fitness. Frequency-independent selection, which is not illustrated here, would not only have a line of zero slope for the fitness-difference function (F_D; that is, $F_a - F_b$), but lines of zero slope for the individual fitnesses of *a* and *b*.

In this discussion, we may seem to have neglected the possible influence of genetic changes that could modify the competing forms, hence the stability point. Ayala (1972), for example, has observed such changes in competing populations of fruit flies under laboratory conditions, when a new genetic form disrupts an established equilibrium and then establishes a new level. The emergence of fitter forms is, of course, crucial in natural selection and in speciation, but, given a mutation, the underlying process determining whether and to what extent it survives appears to be no different from the selection processes already described and

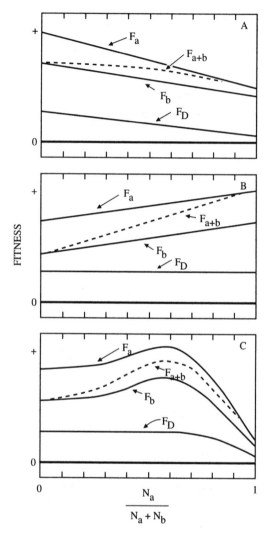

Figure 10.7 Fitness for form a (F_a), for b (F_b), average total fitness (F_{a+b}) and the differences between fitnesses of a and b (F_D), as a function of proportion of a's. In all three panels, b would be excluded by a, since the fitness-difference functions (F_D) are always positive. Exclusion by a minimizes overall fitness (F_{a+b}) for panels A and C, and maximizes it for panel B.

diagrammed in Figures 10.6 and 10.7. Whether the modified form excludes its competition or establishes a new stable mix at a different frequency depends on whether the fitness-difference function passes through zero with negative slope. If the new form has a lower average fitness than that at the prevailing equilibrium point, it simply disappears, which is the case for most mutations.

Another way suboptimal ESS may be superseded has been described in detail by Wilson (1980). Instead of modeling natural selection as if it took place within a homogeneous medium in which the genetic composition (including the genetic variance) is uniform across all localities, Wilson argues for the relevance of what he calls "structured demes." Competition often occurs within localities with varying genetic compositions. The localities may differ in the equilibria they establish for gene frequencies, but they may also differ in how productive they are overall. The localities may be mutually isolated to varying degrees. In such a model, genes compete locally, but localities themselves also compete as they send out colonizers in proportion to their own fecundities. It has been shown formally (if not empirically) that the notion of structured demes, under suitable circumstances, allows a population as a whole to move toward increasing reproductive fitness. The process appears to be the evolutionary analogue to learning new configurations of movements for an individual organism. The melioration process, in either case, operating at a more inclusive level of synthesis, drives the system toward genuine optimization. It remains to be shown how or where the concept of structured demes is relevant to natural selection.

We have been ignoring the rates at which equilibrium is approached by various systems, focusing instead only on the direction in which a system moves. Under some conditions, however, the rates of change can have strong effects. Consider a population of all doves, with an average fitness of 15, and assume it is competing with a second population of animals having an average (and stable) fitness of 10. Two processes will then occur simultaneously: the dove population will gain hawks (by mutation) and approach an average fitness of $6\frac{1}{4}$ (see Figure 10.5); and, of the two competing populations, the one with higher fitness will tend to competitively exclude the other. If the hawk-dove population approaches equilibrium quickly, but competitive exclusion occurs slowly, the hawk-dove population will go extinct (since it, with a fitness of $6\frac{1}{4}$, competes with a population with a fitness of 10). If the two processes are reversed in terms of speed, the other population will go extinct (since it, with a stable fitness of 10, faces a population with a fitness of nearly 15).

Our general conclusion with regard to ESS versus optimization is largely consistent with Dawkins's (1980). In the case of frequency-dependent selection, an ESS analysis or something similar is clearly required, both by existing data and by a logical analysis of the selective process. In the case of frequency-independent selection, an ESS and an optimality analysis lead to the same outcome. But just because of this ambiguity (recall the concurrent VR VR experiment mentioned earlier), cases involving frequency-independent selection cannot tell us whether the process of evolution acts to maximize average fitness, or whether fitter forms displace less fit forms, without regard to maximization. Frequency-dependent selection provides, in effect, an experimental test of the two hypotheses, and it favors ESS. Dawkins concludes that "it seems parsimonious to abandon the phrase 'optimal strategy' altogether, and replace it with ESS" (p. 357). In our view, an ESS analysis is preferable not so much because it is parsimonious, but because it is more nearly *correct*.

The Generality of Stability Analysis

Students of behavioral analysis may recognize that the foregoing analysis of evolutionary equilibrium directly parallels the melioration analysis of behavioral equilibrium. Figure 10.8 shows reinforcement-rate differences between alternative responses in two settings as a function of the time allocated to one of them (see also Prelec, 1983a, ch. 2, Figure 3). Panel A plots the hypothetical schedule shown in Figure 10.4 (B), for which alternative a always reinforces at a higher rate than alternative b, but overall rate of reinforcement varies directly with time allocated to b. The rates or other qualities of reinforcement for a pair of alternatives would always differ by a constant amount ($R_a/T_a - R_b/T_b$ is positive and constant), but the overall level of reinforcement would be an increasing linear function of the time spent on the less reinforcing alternative (alternative b in Figure 10.8). As indicated, melioration ("M") here calls for exclusive choice of a, notwithstanding the optimality ("O") of choosing b exclusively. In evolutionary terms, this would be a case of suboptimal competitive exclusion.

Panels B and C illustrate what, in Ayala's terms, would be called a stable mix of competing (response) forms. Rates of reinforcement for each of a pair of alternatives depended on how much time the subjects (pigeons) were spending at the alternatives. For example, in Panel B it can be seen that from 0–.125 spent at alternative a, the reinforcement rate

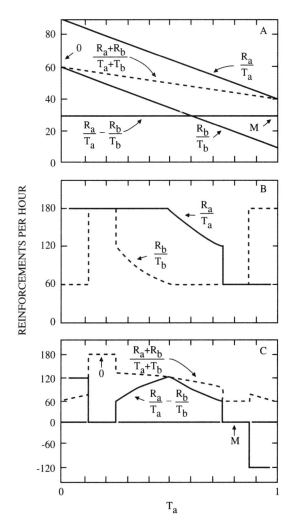

Figure 10.8 A. Overall rate of reinforcement $((R_a + R_b)/(T_a + T_b))$ grows with allocation to b, but a produces a higher rate of reinforcement than b at all allocations. Value is maximized (0) by exclusive preference for b, but melioration (M) predicts exclusive preference for a. This is the hypothetical experiment graphed in Figure 10.4, Panel B. B. Vaughan's procedure (1981) for distinguishing between maximization and melioration. Local rates of reinforcement for a (R_a/T_a) and b (R_b/T_b) are shown as a function of proportion of time spent at a (T_a). C. Overall rate of reinforcement $((R_a + R_b)/(T_a + T_b))$ and differences between local rates of reinforcement $(R_a/T_a - R_b/T_b)$. Rather than maximizing (flat region shown by 0) subjects meliorated (region shown by M).

was higher at a (R_a/T_a) than b (R_b/T_b), but from .875–1.0 spent at alternative a, reinforcement rate was higher at b. The function $R_a/T_a - R_b/T_b$ in Panel C shows reinforcement rate differences at all allocations of time, while the function $(R_a + R_b)/(T_a + T_b)$ shows overall reinforcement rate. This complex function was examined by Vaughan (1981) because it simultaneously provided two tests of the melioration concept. First, it sharply separated the melioration (M) from the optimization (O) equilibrium (see Panel C). Secondly, it provided two points (actually, regions) of intersection with equality in reinforcement rate, only one of which had the negative slope required for stable mixed equilibria. As melioration predicts, pigeons exposed to this schedule all stabilized in the region labeled M, at a substantial cost in reinforcement. Their initial preference has been set at the optimal level, labeled 0, so the shift to M was all the more impressive confirmation of melioration. In this experiment, reinforcement could have been maximized simply by not changing behavior, yet the pigeons changed anyway, as melioration dictated.

The correspondence between maximization of fitness and maximization of utility (i.e., reinforcement) on the one hand, and ESS and melioration, on the other, should now be evident. Evolution and behavioral allocation are analogous processes, because both arise from selection among competing forms (Herrnstein and Vaughan, 1984). Within evolution, a given species does not survive because individual members work for the sake of a common good (e.g., maximization of reproductive fitness). Rather, forms with greater reproductive fitness displace those with lesser fitness, even at a cost in average fitness. Similarly, a given behavior does not survive because it serves a common goal (e.g., maximization of satisfaction). Rather, if one behavior earns more reinforcement than another, allocation shifts toward the first, even if overall reinforcement decreases. At stability, the remaining forms match in average fitness or in reinforcement. If a new form (of genotype or response) arises and survives, it is likely to find a new stability point at a higher fitness or higher reinforcement, because a new form is unlikely to survive unless it does. But, as the hypothetical examples have proved, higher stability points (either for behavior or genes) can in principle cause a reduction in overall reinforcement or fitness, depending on the underlying contingencies.

In evolutionary biology, the underlying contingencies of reproductive fitness are inherent in the interactions among species and their environments. Needless to say, the interactions can be so obscure that great scientific effort is necessary to uncover them. The all too common examples of disrupted ecology owing to the presumably innocuous introduction or removal of a species testify to the subtleties of determining in advance where a new ESS is to be established (e.g., May, 1983).

In contrast, in behavioral analysis, the contingencies of reinforcement are usually chosen in advance by the experimenter. What evolutionary biologists must discover painstakingly, experimental psychologists simply program into computers that run their experiments. This control over contingencies no doubt helps explain why behavioral analysts have (or think they have) discovered the invariance governing behavioral stability, and also why biologists often find psychological experiments too artificial to be applied to species in nature. Nevertheless, because the behavioral contingencies are transparent, it is relatively easy to see that stability is characterized by equal reinforcement rates for the competing responses (the matching law). Although it is not particularly easy to see that the same formal principle determines biological stability, replacing reinforcement with fitness, that has nevertheless been the conclusion of many theorists.

For a time, behavioral biologists believed that the optimizing assumption was ideal for them, because of a notion about evolution. Davies and Krebs (1978), for example, argued for behavior being optimal because it derived from evolution, and evolution tended toward optimal forms. But if natural selection among genotypes tends toward stability regardless of optimality, a different implication should be drawn. In light of the new evolutionary outlook, the argument by analogy at this point is that behavior, too, should tend toward stability, rather than optimality. Dawkins and Krebs (1979) discuss cases in which natural selection appears to lead species to extinction. Similarly, there are many examples of behavioral adaptations turning against the animals that use them.

Consider an experiment on fly larvae reported by Loeb in 1890 (described by Fraenkel and Gunn, 1940). In moving about a larva swings its head from side to side. If one side is darker than the other, it turns its body slightly in that direction as it moves, and continues the swinging motion. By means of this process, it tends to move away from the light in natural situations, presumably an adaptive behavior for a vulnerable and meaty larva.

Loeb placed a number of larvae in a test tube situated at a right angle to a window where, because of the placement of a screen, the part of the tube near the window received dim light. The part of the tube away from the window received bright light. The window was the source of the incident light throughout the tube. By sampling in various directions, the larvae oriented themselves away from the window. Most of them crossed over into the bright light, and continued to move away from the light. By moving away from the window they ended up in the bright end of the test tube. What is adaptive in most natural settings was maladaptive here, but the larvae behaved in their usual way.

Pigeons, presumably, are more advanced than lima beans or fly larvae. Vaughan (1981), in the experiment described above, subjected pigeons to a schedule in which one alternative was reinforced more than another. However, distributing more time to the better alternative reduced the reinforcement from both alternatives. As we have suggested, the pigeons were apparently insensitive to the optimal strategy, and instead meliorated, choosing the locally better, but ultimately more costly, alternative.

Do people do any better? Hardin (1968) (discussed above) suggests that we should not assume that they do. Overexploitation of the environment is a case in which the more immediate reinforcers of indulgence outweigh the delayed reinforcers of forbearance and the long-term consequence is disastrous. On an abstract level, the response of fly larvae, pigeon, and man appear identical. We may conceptualize it as follows. Two situations are sampled (the fly larva samples two light intensities, the pigeon samples reinforcement rates on two keys, the man samples polluting versus not doing so). In all cases, there is a shift in the direction of the contingent payoff: the larva orients away from the window, the pigeon shifts toward the locally better alternative, the man pollutes. In all cases, the long-term effect is maladaptive: the larva ends up in the brighter light, the pigeon receives a lower rate of reinforcement, man degrades his environment.

If man differs at all, it is that knowledge of long-term consequences favors setting up new contingencies of reinforcement (Chapter 4; Ainslie, 1975). For example, a tax or fine for polluting may be sufficient to reduce its reinforcement to the point where it ceases to have a net negative effect on the environment at large. The shift may be viewed as a case of optimization, but it is still an example of melioration, as Figure 10.9 shows. Without the fine or tax, polluting is more reinforcing than not polluting; with a penalty, which the figure depicts as graded beyond some level of indulgence, pollution is held to any arbitrarily low level. A public policy like this may be viewed as a change in reinforcement schedules that exploits melioration to drive behavior toward what is taken to be optimization. In fact, the difficulty people have in creating such presumably rational contingencies of reinforcement, for either their own behavior or for society as a whole, is further evidence for melioration as the controlling process for even human behavior.

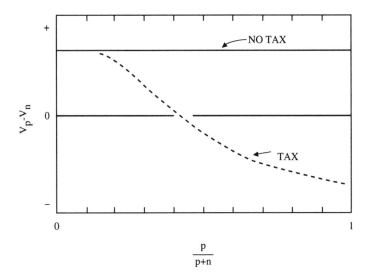

Figure 10.9 Difference between value of polluting (V_p) and value of nonpolluting (V_n), as a function of proportion of industrial residue that constitutes pollution. Without a tax for polluting, the potential polluter always earns more reinforcement for polluting than for not polluting, as illustrated by the solid line. A graduated tax for polluting (e.g., proportion of polluted versus nonpolluted acres of land or cubic yards of effluent) produces the dashed curve. Melioration predicts that pollution will shift from 1 to the abscissa proportion corresponding to the point of intersection with zero on the ordinate.

Where does economics fit here? In fact, economists typically assume that, among competing behaviors (i.e., competing commodities), a stable mix *must* reflect some sort of diminishing marginal substitutability, allowing individual optimization to be preserved. The underlying contingencies of utility are neither discovered (as in biology), nor programmed (as in behavioral analysis), with only rare exceptions. Instead, having assumed that the consumer optimized, economists derive what the utility functions must have been.

Though assuming *individual* optimization, economists have often noted that groups may nevertheless fail to optimize collective (or average) outcomes. The field of welfare economics (e.g., Bohm, 1973) is concerned to delineate, to the extent logically possible, the concept of a social welfare function: How is aggregate welfare affected by changes in individual consumption? However, because interpersonal comparisons of utility are meaningless (a corollary of ordinal, rather than cardinal, utility functions), the question admits of no single answer. Under the usual assumptions of economics, there are, rather, an infinity of Pareto-optimal points, a state of market such that no person could do any better without someone else being made worse off. In spite of the plethora of points of optimization, a violation of Pareto-optimality may be easily constructed.

In the classical "prisoner's dilemma," both prisoners must choose an unpreferred alternative in order to maximize total gain, which is yet one more example of frequency-dependent outcomes. If either or both prisoners tried to maximize individually, the total outcome would suffer (and Pareto-optimality violated). Schelling's (1978) generalization of the dilemma to multi-person groups is, in fact, formally equivalent to our characterization of individual behavior or the ESS analysis of natural selection. If almost everybody carries a tow cable in his car, for example, then it is probably better not to carry one, since it costs something and the other car needed to tow you in any event probably has one. If almost nobody carries a tow cable, it is no doubt advantageous to buy one. The equilibrium state for tow cables versus no tow cables may or may not be optimal for drivers as a whole, depending on the contingencies (and disregarding the incommensurability of individual utilities). Schelling's analysis assumes that individual participants simply choose the better alternative at any given proportion of tow cables, as individual maximization would imply. In fact, individuals themselves do not generally maximize unless, under the contingencies of reinforcement, maximization and melioration converge. For the tow-cable example, at any given frequency of tow-cables, the choice confronting an individual is a binary choice between a larger and a smaller reinforcement, for which both melioration and maximization give the same answer. It is only when the competition among an individual's own responses has the kinds of frequency-dependent contingencies discussed earlier that maximization at the individual level breaks down, just as collective outcomes may be suboptimal if the individual contingencies are frequency dependent (as in the prisoner's dilemma or the tow-cable example).

Groups of individuals may form coalitions that circumvent suboptimal equilibria. They may, for example, form a tow-cable association which identifies its members by stickers on the cars' rear windows. The members promise to help only each other with towing. Similarly, members of labor unions forgo the possibility of improving their salaries by individual effort and negotiation, and rely on collective bargaining instead. In every facet of social life there may be risks in the unrestrained pursuit of advantage. The social contract itself replaces the Hobbesian confrontation of all against all. Like the structured demes of population biology, or the restructured response categories of individual behavior, people sort themselves into aggregations, which then interact. How and why these groups arise are questions that lie outside our concern here, except to note that they complete the parallel we have tried to draw at three levels of analysis.

At the genetic, behavioral, or collective levels, elements selected by individual consequences achieve equilibria that may or may not maximize the variable involved in selection. Individual genes selected for fitness may or may not maximize total fitness; responses strengthened by reinforcement may or may not maximize total reinforcement; participants recruited by outcomes may or may not maximize a group's collective outcome. A stable equilibrium involving mixtures rather than total exclusion—of genes, responses, or participants—must always obey a matching law, an equality (assuming commensurability) in the ratio of the frequency of each element to its selective principle: genes to their fitness, responses to their reinforcement, participants to their outcomes.

Summary

The assumption that the behavior of organisms conforms to principles of optimization has derived strength from two distinct sources: the rationalistic perspective, absorbed by economic theory in the nineteenth century (people act to maximize satisfaction or utility), and the maximization of fitness assumption within evolution (a maximally fit organism will, as a corollary, act to maximize rate of reinforcement). Thus, a deduction from the evolutionary argument serves to complement an earlier argument for rational behavior.

Our position, first, is that behavior and evolution each exemplify a single principle of variation and selection, and, second, the principle can lead to equilibria that need not be optimal. Data that have previously been said to verify optimization are, we suggest, actually ambiguous with regard to distinguishing between optimization and a principle of melioration. And, given situations that allow one to discriminate between the two processes, the data unequivocally favor melioration.

11 | Rational Choice Theory: Necessary but Not Sufficient

We start with a paradox, which is that the economic theory of rational choice (also called optimal choice theory) accounts only poorly for actual behavior, yet it comes close to serving as the fundamental principle of the behavioral sciences. No other well-articulated theory of behavior commands so large a following in so wide a range of disciplines. I will try to explain the paradox and to present an alternative theory. The theory of rational choice, I conclude, is normatively useful but is fundamentally deficient as an account of behavior.

Rational choice theory holds that the choices a person (or other animal) makes tend to maximize total utility, where utility is synonymous with the modern concept of reinforcement in behavioral psychology. Because utility (or reinforcement) cannot be directly observed, it must be inferred from behavior, namely, from those choices themselves. Rational choice theory is thus a rule for inferring utility: It says that what organisms are doing when they behave is maximizing utility, subject to certain constraints. Rational choice theory is also used normatively, as a way of assessing whether behavior is, in fact, optimally gaining specified ends, and if not, how it should be changed to do so. The distinction between

Originally published in *American Psychologist*, 1990, 45, 356–367. Copyright © 1990 by the American Psychological Association. Reprinted with permission. The author gratefully acknowledges the large contributions to this work of William Vaughan, Jr., Dražen Prelec, Peter de Villiers, Gene Heyman, George Ainslie, James Mazur, Howard Rachlin, and William Baum. Interested persons may find more precision and detailed accounts of the data in Chapter 10; Herrnstein 1970, 1988; and, especially, Williams, 1988. Several anonymous reviewers and associate editor Donald Foss deserve thanks for uncommonly helpful comments on an earlier version. Thanks are owed, as well, to the Russell Sage Foundation for support and an environment during 1988–89 that provided an opportunity for a study of the relations between economic and psychological theories of individual behavior.

descriptive and normative versions of rational choice theory is fundamental to the theme of this essay.

The theory of rational choice seems to stand in relation to the behavioral sciences as the Newtonian theory of matter in motion stands to the physical sciences. It is held, by its proponents, to be the law that behavior would obey if it were not for various disruptive influences, the behavioral analogues of friction, wind, measurement error, and the like.

Not just economics, but all the disciplines dealing with behavior, from political philosophy to behavioral biology, rely increasingly on the idea that humans and other organisms tend to maximize utility, as formalized in modern economic theory. In accounts of governmental decision making, foraging by animals, the behavior of individual or collective economic agents, of social institutions like the criminal justice system or the family, or of rats or pigeons in the behavior laboratory, it has been argued forcefully that the data fit the theory of rational choice, except for certain limitations and errors to which flesh is heir. The scattered dissenters to the theory are often viewed as just that—scattered and mere dissenters to an orthodoxy almost as entrenched as a religious dogma.

How can anyone plausibly subscribe to the descriptive theory of rational choice in the face of the reality that organisms often behave against self-interest? Even some rational choice theorists procrastinate and suffer from other human frailties. They may overeat, smoke, drink too much, and make unwise investments, just like the rest of us. People may behave altruistically at some personal sacrifice. Martyrs are just rare, not unknown. Neither the existence of unwise nor altruistic actions evidently wounds the descriptive theory of rational choice for its most committed adherents.

A resistance to ostensibly contrary data is not unique to rational choice theory. It has often been observed that scientific theories evolve to cushion themselves from the hard knocks of data; neither rational choice theory nor the alternative theory to be proposed here is an exception to this generalization.

But that general resistance to counterevidence is not the only reason rational choice theory endures. Behavior that might seem irrational because it is not guided by obvious self-interest is sometimes explained in rational choice theory by invoking whatever source of utility is needed to rationalize the observed behavior. This is possible within the theory because utility, which is subjective, differs from objective value. There is, in principle, no constraint on utility other than that imposed by the behavior from which it is inferred. In principle, nothing prevents inferring utilities that lead to self-damaging or altruistic behavior, for example. A similar stratagem is available to reinforcement theorists, who are also free to infer reinforcement from the observed behavior.

We may, for example, be optimizing subjective utility (or reinforcement) by eating ice cream and red meat and smoking dope, even though we are, and know we are, harming ourselves. Some people give up a great deal, objectively speaking, for the subjective utilities they are presumably deriving from cocaine or alcohol, including shortening their lives and decreasing the quality of their lives. The things that organisms strive to obtain or to eliminate are taken as givens by the theory. When rational choice theorists say, "De gustibus non est disputandum," they mean it (Stigler and Becker, 1977). Rationality, in this modern version, concerns only revealed preference.

Not only are utilities subjective, says the theory of rational choice, but so are the probabilities by which they can be discounted by uncertainty. People often act as if they overestimate low, but nonzero, probability outcomes and underestimate high probability outcomes, short of certainty. They may worry too much about, and pay too much to insure themselves against, low-probability events such as airplane accidents. People insure their cars against improbable losses, then, with abandon, run red lights on heavily traveled city streets. After working hard to earn their pay, they buy lottery tickets with infinitesimal odds of winning. Instead of objective probabilities, it has been proposed that utility theory must take into account subjective weights, bearing complex, as yet unexplained, relations to objective frequencies.

The subjectivity of utility is motivational. The subjectivity of probability is cognitive. Rational choice theorists invoke other psychological complications beyond these, having to do with limitations in organisms' time horizons, knowledge, capacities for understanding complexity, and so on. Acknowledging those limitations, while saving the theory, is like the postulation of epicycles in planetary astronomy, in either case smoothing the bumpy road between facts and theory. The question is whether the epicycles of rational choice theory are protecting a theory that inhibits understanding or advances it, whether the correct analogy is Ptolemy's geocentric theory or Copernicus's heliocentric theory, each with its own epicycles.

As a descriptive theory, rational choice theory survives the counterevidence by placing essentially no limit of implausibility or inconsistency on its inferred utilities and also by appealing to the undeniable fact that organisms may calculate incorrectly, be ignorant, forget, have limited time horizons, and so on. Other lapses of rationality, as they are illuminated by the numerous ingenious paradoxes of choice research, are often swiftly absorbed by the doctrine of rational choice, at least in the eyes of its most devoted followers. Those odd, obscure, or shifting motives and those errors of calculation and time perspective aside, we are all rational calculators, the theory says.

Rational choice theory also survives because it has several genuine strengths, beyond its indisputable value in normative applications. First, rationality accords with common sense in certain simple settings. For example, consider a choice between $5 and $10, no strings attached. Any theory of behavior must come up with the right answer here, where there seems to be no issue of obscure motives, or of errors of reckoning, remembering, knowing, and so on. Assuming only that more money has more utility than less money, rational choice theory does come up with it. To argue against rationality as a fundamental behavioral principle seems to be arguing against self-evident truth.

Second, rational choice theorists have formalized utility maximization, reducing it to its axiomatic foundations. Many of the most brilliant theoreticians are drawn to this part of the behavioral and social sciences, for here is where their powerful intellects shine most brightly, addressing questions of formal structure, not distracted by the fuzziness of motivation or the messiness of data. Some rational choice theorists admit that the theory is wrong, but they see no good reason to give up something so elegantly worked out in the absence of a better theory. Many rational choice theorists evidently believe that no theory could simultaneously describe behavior better than, and be as rigorous as, rational choice theory. Real behavior, they seem to believe, is too chaotic to be rigorously accounted for with any precision.

The foundations of rational choice theory have, however, lately been under attack. Experimental findings by many decision researchers (e.g., Kahneman, Slovic, and Tversky, 1982) have undermined the descriptive form of the theory by discovering choice phenomena that are consistent with (or at least not inconsistent with) principles of cognitive psychology, but inconsistent with rationality as commonly construed. Bombarded by these data, the unifying concept of rational choice may give way to a set of psychological principles, none of which is of comparable breadth, but which, in the aggregate, will account for actual behavior better than the global assumption of rationality (an approach exemplified in a recent textbook by Dawes, 1988).

Theoretical challenges also abound. It has been repeatedly suggested that it is not individual behavior that satisfies principles of rationality, but natural selection (e.g., Frank, 1988; Hirshleifer, 1982; Margolis, 1987). Evolution, guided by natural selection, endows individuals with behavioral rules of thumb that may be individually suboptimal, but that in the aggregate, approximate optimality in some sense (Heiner, 1983; Houston and McNamara, 1988). A few theoreticians (e.g., Luce, 1988, 1989; Machina, 1987), drawing mainly on the paradoxes of choice in the face of uncertainty (e.g., the familiar Ellsberg and Allais paradoxes, discussed in Dawes, 1988), have been exploring the possibility of relaxing one or another of the axioms of rationality while retaining the rest of the formal theory.

At least a few (and perhaps many) economists and other social scientists would, at this point, defend rational choice theory only in its normative form and would agree that the descriptive form has lost its credibility in the face of too many "anomalies" of individual behavior—too many epicycles, in other words. For many of these theorists, there is a theoretical vacuum as yet unfilled. One can predict a surge of new theories to fill the void. In this article, I will attempt to fill a part, if not all, of the vacuum with a theory arising out of the experimental analysis of behavior.

The advantages of the present theoretical alternative are that it accords no less well with common sense than rational choice theory, that it lends itself to as rigorous a formal structure, that it has extensive empirical support, and that it is consistent with many of the irrational behaviors we actually observe in ourselves and others. The primary disadvantage, which may or may not prove to be decisive, is that the large experimental literature on which it is based comes mainly, though not exclusively, from studies of animal rather than human subjects.

Some Systematic Irrationalities

The weaknesses in rational choice theory are uncovered by systematic inconsistencies in behavior, which can sometimes be graphically illustrated by asking people to solve riddles. Their solutions may betray the inconsistencies. I will consider two riddles and one experiment that point toward the alternative theory to be developed here. However, even in advance of an account of the theory I am proposing, the riddles and the experiment show that something goes wrong when people are asked to make certain kinds of choices.

Suppose a person is asked to imagine winning a lottery and is given a choice between $100 tomorrow and $115 a week from tomorrow.[1] Whichever the person chooses (only hypothetically, because no money is given), the money is said to be kept in escrow by a Federal Reserve bank, then delivered by bonded courier. Now the person is asked to choose one. When I present a problem like this, a fair proportion of people choose the earlier but smaller payoff.

Now, those who choose the small payoff are asked to imagine winning another lottery and are given a choice between $100 tomorrow and $140,000 a year from tomorrow. Again, the Federal Reserve holds the money and delivers it on the schedule chosen. Everyone, I find, picks the more deferred but larger prize.

Finally, consider winning yet another lottery. The person is asked to choose between $100, 52 weeks from today or $115, 53 weeks from today. The Federal Reserve will do its usual fine job of holding and delivering the money. Most of the people who chose $100 in the first lottery switch to $115 here.

This natural pattern of choices violates the consistency implicit in rationality, and it does not seem to be a matter of obscure motives or of incidentally faulty arithmetic. Some more fundamental flaw in our decision making appears to be responsible. In the first lottery, those who choose $100 have, by their choice, revealed a discount rate of more than 15% per week. They have, in effect, said that they would be willing to forgo $15 (possibly even more) to get $100 a week sooner. If their discount rate was less than 15%, they would have chosen the later $115 over the earlier $100.

In the second lottery, the choice of $140,000 reveals a discount rate smaller than 15% per week, because when $140,000 is discounted at 15% a week for 52 weeks, the result is $97.69, less than the $100 the person could get by choosing the earlier payoff. As odd as it may seem, someone who thinks $100 tomorrow looks better than $115 deferred

---

1. A version of the riddle using $100 and $120 was described by Herrnstein and Mazur (1987). No formal experiment has been done with either that version or the present one, but from informal observations, it is clear that many people succumb to the inconsistency described here. The quantitative features of the inconsistency have not been explored under controlled experimental conditions.

for a week should also think it looks better than $140,000 deferred for a year, if rationality prevailed.

From past experience, I know that some people, confronted with this lack of consistency in their choices, staunchly defend their rationality. They say things like, "I chose the smaller amount in the first lottery because another $15 isn't worth my thinking and worrying about for an extra week. An extra $139,900, however, is another matter altogether, well worth waiting a year for." It is because of such excuses that we add the third lottery, because here, too, one would be thinking about collecting another $15 for an extra week, yet most people find it worthwhile to do so when the week is a year deferred.

Nothing in rational choice theory can explain this curious inconsistency, yet it seems to be an example of an almost ubiquitous tendency to be overinfluenced by imminent events. The tendency toward impulsive, temporally myopic, decision making causes considerable grief, as we all know. Let us be clear about how the example exemplifies irrationality. The mere discounting of deferred consequences need not be irrational. If one postpones payment for work done or goods delivered, one will have to pay more than if one pays immediately. The sellers may calculate rationally that they are forgoing interest they could be earning or pleasure that they could be harvesting while the buyers hang on to the payment and garner the fun or the interest. Even if they are not calculating human beings, but rats or pigeons in a behavioral experiment, deferred consequences are likewise downgraded. Perhaps natural selection has already factored in something functionally equivalent to the rational consideration of foregone benefits.

In either case, if the discounting is rational, the rate should be fixed per unit time, barring gratuitous assumptions. Fifteen percent a week is 15% a week, now or next year, in the theory of rational choice. In the example, however, we reveal that we downgrade not only value, but also the rate at which we downgrade value. The discount rate may be 15% for next week, but for a week a year from now, the discount rate itself has shrunk so much that it leaves $115 looking better than $100 even though they are separated by a week.

Many problems of choice spread over time have a similar shape. Imagine, for example, that we could always select meals for tomorrow, rather than for right now. Would we not all eat better than we do? We may find it possible to forgo tomorrow's chocolate cake or second helping of pasta or third martini, but not the one at hand. People who are trying to lose weight pay dearly to spend time in dieting resorts ("fat farms"), where what they get for their money is losing the option of not eating on their own. The examples reveal our tendency to be inconsistent because of impulsiveness.

A poignant example of temporal myopia is provided by the discovery of genetic markers for Huntington's disease, a progressive, fatal disease of the nervous system. The disease is typically asymptomatic until early adulthood or middle age. It is caused by a single, dominant gene, so that an offspring of one parent with the disease faces a 50–50 chance of having it himself or herself. It is now possible for people facing this risk to find out early in life, with high accuracy, whether or not they carry the gene.

By far, most of the people at risk have declined to take the test (Brody, 1988). This reluctance would make sense within a rationalistic framework if it were the case that the negative subjective change from a 50–50 chance to a virtual certainty of the disease were larger than the positive subjective change from a 50–50 chance to a virtual certainty of no disease. That, however, is the reverse of the evidence described in the newspaper article just cited.

People who know they face an even chance of this fatal disease have typically already factored much of the worst possible news into their lives, by choices made about marriage, parenthood, occupation, and so on. If their fears are confirmed, there is an increment of sorrow, a resignation to a fate already played out in their minds, but no huge change in subjective state. The newspaper account says that bad news triggers no visible increment in psychopathology or need for tranquilizers. In contrast, those who get good news experience enormous joy and relief. Over time, their lives probably readjust to normality. But even given this dramatic asymmetry favoring positive subjective change over negative, few people take the test.

The Huntington's example is faintly echoed in what happens when we stand in water up to our knees at the beach on a hot day, knowing that relief is only a few moments away if we plunge in.[2] But, instead, we are daunted by anticipation of those icy first few seconds. It can be so hard to overcome this barrier that we give up and turn back to the hot beach. Sometime between when we first left the blanket on the beach and when we hesitate knee deep, the promise of relief has been swamped by the avoidance of the rapid drop of temperature.

---

2. I owe this comparison to George F. Loewenstein.

Note that these examples resemble the lotteries described earlier, in that an immediate consequence (e.g., the pleasures of food, a 50% chance of an increment of sorrow from a negative test, or avoiding the icy plunge) is chosen over a deferred alternative (weight loss, a 50% chance of life free from the threat of Huntington's disease, or cool relief). Moving the consequences of choice away from the present, while holding constant everything else about them, often reverses the preference order. For eating and for taking the plunge, it is plain that the preference reverses. For Huntington's disease, we can surmise that it also reverses, because most of us would advise a person at risk to take the test (as physicians now do advise them), but are likely to be unable to do so when we face the prospect of immediate bad news ourselves.

In each case, the discounting factor applied to restraint in relation to impulse shrinks as it moves further in time, so we choose impulsively when the consequences are at hand, but with restraint when they are deferred. We are disposed to see things in better perspective as they become more remote. How come?

One approach is to invoke a systematic psychophysical distortion of time perception, foreshortening remote time intervals. That may, indeed, be true, but an answer[3] closer to the data and of more fundamental significance is that we discount events hyperbolically in time (at least approximately: see Chapter 5; Ainslie, 1975; Mazur, 1985, 1987; Williams, 1988), rather than exponentially, as rational choice theory assumes. A hyperbolic time discounting function has, as one of its corollaries, the very foreshortening of remote time intervals that the data suggest. With exponential discounting, the discount rate remains fixed; with hyperbolic, the rate itself shrinks with time.

---

3. The answer is contemporary, but the question of time perspective in choice is not. I thank James Q. Wilson for calling my attention to David Hume's characterization of it in the 18th century, from an essay on the origins of government: "When we consider any objects at a distance, all their minute distinctions vanish, and we always give the preference to whatever is in itself preferable, without considering its situation and circumstances. . . . In reflecting on any action which I am to perform a twelvemonth hence, I always resolve to prefer the greater good, whether at that time it will be more contiguous or remote; nor does any difference in that particular make a difference in my present intentions and resolutions. My distance from the final determination makes all those minute differences vanish, nor am I affected by any thing but the general and more discernible qualities of good and evil. But on my nearer approach, those circumstances which I at first overlooked begin to appear, and have an influence on my conduct and affections. A new inclination to the present good springs up, and makes it difficult for me to adhere to my first purpose and resolution. This natural infirmity I may very much regret, and I may endeavor, by all possible means, to free myself from it" (Hume, 1777/1826, pp. 314–315).

Exponential time discounting arises from rationalistic considerations; hyperbolic time discounting is a frequent result of behavioral experiments on various species, including human. The evidence for hyperbolic discounting comes primarily from choice experiments in which it is assumed that the subjects are obeying the matching law, a principle of choice that has been widely observed in the laboratory and is defined here in the context of the next riddle to be discussed (Chapters 5, 6, 7; Ainslie, 1975).

Imagine that a person is playing tennis, and her or his opponent comes to the net (Herrnstein, 1989; Herrnstein and Mazur, 1987). Assume that the person must now choose between a lob and a passing shot and disregard, for simplicity, any strategic plan in which the opponent may be engaging. Consider the opponent a random variable. Both lobs and passing shots are more effective if they are surprising, and less effective if they are expected. Assume, finally, that surprise has a larger effect on the effectiveness of lobs than of passing shots, which is probably the case. How does he or she decide which shot to hit?

I have presented this riddle to many people, including devotees of rational choice theory. Almost everyone who agrees to play along comes up with something like the following: "As long as one shot is more effective than the other, I'd use it. As I use it, the surprise factor takes its toll. When the other shot becomes more effective, I'd switch to that one. And so I'd oscillate from one shot to the other, trying to switch to the one that is currently more effective."

No one to whom I have presented the riddle has ever spontaneously noticed that the strategy I just characterized may be significantly suboptimal. Some concrete values may help. Suppose the lob has a .9 chance of earning a point when it is a surprise and a .1 chance of doing so when it is fully expected. A surprise passing shot, we can assume, has a .4 chance of being effective, and a .3 chance if it is fully expected. Figure 11.1 plots these points and connects them linearly for intermediate levels of expectation, as functions of the expectation for a lob. The dashed curve is the joint effect of both shots, which is to say, the average of their effectiveness weighted by the relative frequency of their use. Figure 11.1 assumes that expectations for the two shots are determined by the probability of their use in the recent past and that the probability of one is the complement of that for the other.

The strategy that people espouse falls at the intersection of the two solid lines in Figure 11.1. It is here, at about two-thirds lobs, that the two shots have equal effectiveness. A shift toward more lob use reduces

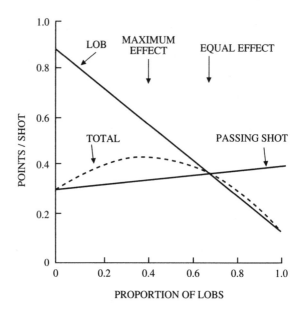

Figure 11.1   Points per shot for a hypothetical tennis player choosing between lobs and passing shots as functions of the current probability of lobs. *Note:* Both shots profit from surprise, but lobs do so more than passing shots. The behavioral equilibrium point is at about two-thirds lobs, but the optimal strategy is at about 40% lobs. Data are from "Darwinism and Behaviorism: Parallels and Intersections" by R. J. Herrnstein. In *Evolution and Its Influence* edited by A. Grafen, 1989, London: Oxford University Press. Copyright 1989 by Oxford University Press. Data are also from "Making Up Our Minds: A New Model of Economic Behavior" by R. J. Herrnstein and J. E. Mazur, 1987, *The Sciences,* November/December. Copyright 1987 by New York Academy of Sciences. Used by permission.

the effectiveness of lobs and likewise for more passing shot use. This is a point of equilibrium in the sense that deviations from it are self-negating, if the player is using the strategy of comparing the effectiveness of the shots.

If the player were a point-maximizer, however, she or he would use a different strategy. The player would look at the two shots overall and pick the point at which their joint effectiveness is at a maximum, shown in Figure 11.1 as the maximum of the dashed curve, near 40% lobs. At the maximum, each lob is more effective than each passing shot, but the two of them together provide the highest returns. Even after I try to explain where the maximum strategy lies, many people express puzzlement. Finding the maximum in a situation like this does not seem to come naturally.

What does come naturally, as noted earlier, is the strategy that stabilizes at the intersection of the two solid lines, where both shots have the same average value in points. This distribution of shots is dictated by the matching law. According to the matching law, behavior is distributed across alternatives so as to equalize the reinforcements per unit of behavior invested in each alternative. Or to put it another way, the proportion of behavior allocated to each alternative tends to match the proportion of reinforcement received from that alternative. The tennis riddle thus provides an example of spontaneous human irrationality and of the relation of that irrationality to the matching law.

In several hundred experiments, mainly on animals but also on human beings, choice has approximately conformed to the matching law (for recent reviews of the literature, see Davison and McCarthy, 1988; Williams, 1988). The complexities of this literature would be out of place here, but simply stated, the widely accepted conclusion is that subjects allocate behavioral alternatives so that each alternative action earns the same rate of reward per unit of behavior invested, once variations in response topography and in reward quality have been taken into account. This equalization of reward rates is the matching law and is exemplified by the allocation people choose in the tennis riddle. Although the matching law is well established, there are varying explanations for its occurrence.

William Vaughan, Dražen Prelec, and I did an experiment (described in Herrnstein and Prelec, 1991) on human subjects that is reminiscent of the tennis riddle.[4] Its results suggest what the dynamic process is that causes people (and animals) to obey the matching law. Volunteer subjects spent a half hour or so in an experimental booth containing two response keys. The subjects were told they would earn a few cents every time they depressed either key when the trial light was illuminated. The probability of reinforcement was, in other words, 1.0 under all conditions. Each trial was separated from the next by an intertrial interval. The intertrial interval following a choice of one of the keys (the 1-key) was two seconds shorter than that following a choice of the other key (2-key). However, the intertrial interval following either choice was an increasing linear function of the proportion of 1-key choices in the preceding 10 trials (including the present one). Figure 11.2 lays out the relations.

---

4. This is a version of "the Harvard Game" described in the introduction to this section. In that version the number of points differed between alternatives. In this version points were equal between alternatives but intertrial intervals differed. [Editors' Note.]

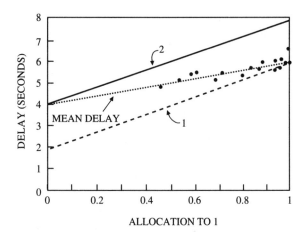

Figure 11.2   Delay between trials following key 1 (dashed line) and key 2 (solid line) choices, as functions of the proportion of key 1 choices in the preceding 10 trials. *Note:* The dotted line is the weighted average of the delays following the two choices, at the prevailing allocation to key 1. A single session's behavior for each of 17 subjects is shown by the dots. Data are from an unpublished experiment by Prelec, Vaughan, and Herrnstein.

Figure 11.2 shows the intertrial interval following 1- and 2-key choices (dashed and solid lines, respectively) as functions of the proportion of 1-key choices during the 10 preceding trials and the weighted average of both delays (dotted line). After the verbal instructions, subjects were given 100 free trials to familiarize themselves with the procedure, then a 10-minute period for playing "for keeps." They were advised that it was in their best interest to complete as many trials as possible in the 10-minute period, but were told nothing more about the contingencies.

The 17 dots show the performances of 17 subjects in a single session. Some subjects worked for extended times over multiple sessions, without any systematic change in their patterns of responding. The optimal strategy would have been to choose key 2 every time, thereby minimizing the intertrial interval at 4 seconds. Instead, all but one of the subjects chose key 1 most of the time, and a few chose it virtually all of the time, enduring the worst possible overall delay between trials, 6 seconds.

Having been subjects ourselves (though not included among the 17 plotted), we think we know what happens. A two-second difference in intertrial intervals is hard to disregard. It therefore feels right to add more key 1 responses into the mix of choices. One strategy that is immune to that temptation is exclusive preference of key 1, which is the worst possible strategy. Subjects sometimes sense that their choices are influencing the intertrial intervals, but they do not know what to make of the information. Likewise, with the tennis riddle, telling people exactly how each shot's effectiveness interacts with its use does not guide them to a proper application of the information. The next section attempts to characterize the underlying process in general terms.

The rational choice advocates I talk to are likely to complain that these two examples—the tennis riddle and the intertrial interval experiment—are too complex for the basic maximizing tendency to emerge. "Give people a chance," they say, "and they will maximize. Of course, if you mix them up badly enough, they won't." That argument calls for two rhetorical rejoinders. One is that if these contingencies are complex enough to suppress our basic maximizing tendency, then it is a fragile tendency indeed. One may argue about how to define simplicity and complexity, but not many decision problems in ordinary life are simpler by any defensible definition of those terms.

The second reply is that it would have been uncharacteristic for natural selection to endow us with a decision rule, such as the maximizing principle, and to fail to endow us with the capacity to exploit the relevant variables. Creatures are usually remarkably sensitive to the stimuli controlling their behavior—like fish responding to the currents in the water around them. With maximization problems, we seem to be like fish out of water.

Melioration and Matching

More to the point, however, is that the results for both of these examples conform to a familiar finding in the behavior laboratory (Chapter 4; Herrnstein and Vaughan, 1980; Prelec, 1982), one that may explain the matching law. Confronted with choices that provide differing rates of reward per unit of behavior invested, organisms allocate more time to the alternatives that provide the higher rates of reinforcement. This tendency to shift behavioral allocation toward more lucrative alternatives has been called melioration.

As a subject meliorates, that is, as it shifts toward better alternatives, its doing so may cause the reinforcement returns from the alternatives to change, for various reasons.[5] Behavior then continues to shift toward the newly better alternatives. We would shift from lobs to passing shots and vice versa as one or the other was doing better on the average. In the intertrial experiment, subjects shift toward key 1 because it is followed by a shorter delay. The principle of melioration seems commonsensical enough; the only surprises may be that it fails to maximize reinforcement and that it leads to an equilibrium dictated by the matching law.

The melioration principle says that choice is driven, in effect, by a comparison of the average returns from the alternatives. Equilibrium is attained either when one alternative has displaced all others or all the remaining alternatives from the choice set are providing equal returns per unit consumption. Either of these equilibrium conditions conforms to the matching law, according to which the relative frequency of each behavioral alternative matches the relative frequency of reinforcement provided by it (Herrnstein and Vaughan, 1980). The relation between the matching law and hyperbolic time discounting is exemplified in Chung and Herrnstein's experiment on delayed reinforcement (see Chapter 5) and in a large variety of procedures explored by Mazur (Mazur 1985, 1987; Mazur, Snyderman, and Coe, 1985; Mazur and Vaughan, 1987; see also the summary in Williams, 1988).

Melioration and matching can produce suboptimalities when one's allocation of choices affects the returns we obtain from the alternatives, as in both the tennis riddle and intertrial experiment (see Chapter 14 for a formal treatment). Such interactions are the rule in everyday life, not the exception. When they take place, the natural decision-making tendency is evidently to disregard the implications of this interaction for overall returns and to focus instead on the current average returns from the alternatives, which is to say, to meliorate, hence to match.

In the tennis riddle, the familiar strategy exemplifies melioration— shift from one shot to the other when the other becomes more effective. In the intertrial experiment, melioration drives choice toward key 1, the alternative with the shorter delay following it, which maximally defies rational choice theory. Comparably stark violations of rational choice theory have been observed in more systematic experiments.

---

5. They may change for motivational or situational reasons. For example, when a food source is chosen more often, hunger may be reduced, making a quantity of food less reinforcing, or the source may be depleted, reducing the average return per unit of behavior invested.

For example, pigeons in an experiment chose between two alternatives, each delivering a bit of food at irregular intervals with average values that were varied parametrically during the course of the study (Heyman and Herrnstein, 1986). For one alternative, the clock that timed the intervals ran only when the pigeon was choosing that alternative; for the other alternative, the clock ran all the time, but the reinforcements it scheduled were only given to the pigeon after it chose that alternative. The pigeon could switch from one alternative to the other at will, with a single peck at a key.

To earn the maximal reinforcement rate, the pigeons should have spent most of their time on the alternative for which the clock ran only when the alternative was chosen, sampling the other alternative occasionally, to collect reinforcements that were due. But maximizing reinforcement here would violate melioration and matching. Numerous workers have lately studied variants of this schedule (known as a concurrent variable-interval, variable-ratio schedule) because it sharply discriminates between the melioration principle and the reinforcement maximizing principle. The evidence to date has clearly, if not unanimously, favored melioration (see reviews in Heyman and Herrnstein, 1986, and Williams, 1988).

A particular transition in one experiment is shown in Figure 11.3. Midway through the experiment, two pigeons had just started working on a new pair of scheduled values. The reinforcement maximization principle predicted that they would spend none of their time on the alternative called VT, the one for which the clock runs continuously. The melioration principle predicted that they would spend almost all their time choosing VT. Because of the schedule parameters in the preceding condition, the pigeons started off spending a quarter or less of their time on VT, as the open points indicate. They started off at a reinforcement rate close to the maximum possible.

However, as the lines between the open and closed circles indicate, they also started off in this new condition quite far from the point predicted by the matching law. Over several weeks of daily sessions, the pigeons gradually shifted in their choices toward the VT alternative, as melioration drove them toward conformity with the matching law. The lines between circles shrank to virtually nothing, indicating conformity with matching. As they did, earnings fell almost steadily, traced by the triangular-shaped points. Obeying the matching law cost the pigeons more than a third of their overall rate of food reinforcement. This condition's results were typical. In the experiment as a whole, the pigeons

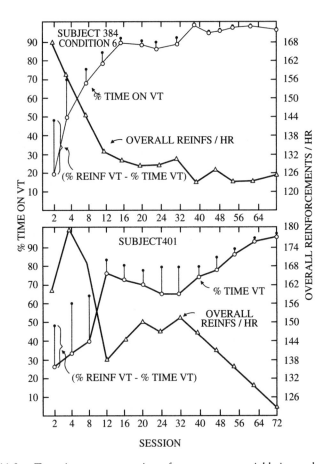

Figure 11.3 Two pigeons on a version of a concurrent variable-interval, variable-ratio schedule of reinforcement. *Note:* Each starts off spending about 20–25 percent of its time on the alternative called VT, shown by the open circles, where the overall rate of reinforcement, shown by the triangular symbols, is near maximal. However, the deviation from the matching point, shown by the solid dots, is also near maximal. Over sessions, behavior shifts toward more time on VT, which reduces the deviation from matching at the cost of lost reinforcement. Data are from "More on Concurrent Interval-Ratio Schedules: A Replication and Review" by G. Heyman and R. J. Herrnstein, 1986, *Journal of the Experimental Analysis of Behavior, 46.* Copyright 1986 by Society for the Experimental Analysis of Behavior. Used by permission.

earned food at a lower rate than they would have by allocating choices randomly to the two alternatives, let alone what they could have earned as food reinforcement maximizers.

Does real life ever arrange anything like this schedule for pigeons or other species? Houston (1986) has shown that a schedule much like it governs food availability for the pied wagtail, a bird living in the Oxford University environs where he and his colleagues work. Field studies confirm that the foraging of the pied wagtail is suboptimal in the way that melioration predicts for this sort of schedule, with too much time being invested in the alternative that needs to be sampled only occasionally to gain maximal food recovery.

The melioration principle implies suboptimality particularly for the class of situations that Prelec and I have characterized as "distributed choice" (see Chapter 14). For distributed choices, the organism does not make a once-and-for-all decision about the alternatives in a choice set. Instead, repeated choices are made over some period of time, and the decision variable is the allocation among alternatives, the ongoing proportion of lobs versus passing shots, for example.

Most "style of life" questions concern distributed choice, Prelec and I have suggested. At no moment in life does one choose, for example, to become promiscuous or a glutton or an alcoholic. Rather, those human frailties creep up on us insidiously, the result of numerous minor choices, many of which may be barely, if at all, blameworthy. Other examples best considered as matters of distributed choice are easy to think of: the continuum from miserliness to spendthriftiness, from profligacy to prudery, or from being an exercise junkie to being sedentary. No single choice is involved in being diligent rather than lazy, honest rather than dishonest, loyal rather than disloyal, clean rather than dirty, and so on.

Rational choice theory has not formulated a clear or effective framework for distributed choice, although some theorists have addressed particular examples, such as drug addiction. The hope has been that the concepts applied to the idealized, timeless choices in their theory will smoothly generalize to temporally extended fields, to distributed choice. Evidently, they have not. In contrast, in the experimental analysis of behavior the study of behavioral allocation within a period of observation has been central for more than a generation. Both melioration and matching describe behavior in the framework of distributed choice.

Figure 11.4, for example, is our account of the suboptimality observed in an experiment such as that just summarized. This is the same sort of chart presented earlier for the intertrial interval experiment. The

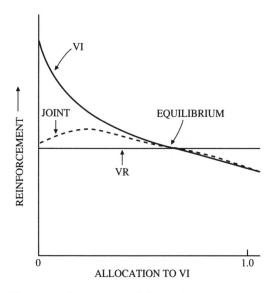

Figure 11.4   Schematic representation of the reinforcement structure of a concurrent variable-interval, variable-ratio schedule. *Note:* The reinforcement rate for the variable interval (VI) is inversely related to the time allocated to VI; for the variable ratio (VR), reinforcement rate is independent of time allocated to either alternative. Consequently, the equilibrium point (i.e., matching) is displaced from the allocation that would maximize reinforcement.

x-axis plots the proportion of time spent on the alternative that needed to be sampled only occasionally because its clock ran continuously (variable interval or VI). As a function of this variable, the y-axis gives the rate of reinforcement received from this alternative while the subject is choosing it. Because its clock runs all the time, the less time the subject spends on it, the higher the rate of reinforcement when it is sampled. It can be thought of as a model of a reinforcement source that gets depleted as it is sampled and restores itself when it is left unsampled, or as a motivational state that fluctuates with deprivation and satiation.

Also shown in Figure 11.4, as a function of time spent on this alternative, is the rate of reinforcement received from the other alternative (variable ratio or VR). Because the other clock runs only when the other alternative is being sampled, it provides a fixed rate of return per unit of time invested in it, however much the subject samples it. It can be thought of as a model of a standard gambling device, with a fixed probability of winning. Finally, Figure 11.4 shows, as a dashed curve, the joint returns from the two alternatives.

When the subject is spending only a small proportion of its time on the VI, then the schedule provides a higher rate of return than the VR. Melioration then dictates increasing time on the VI. However, if the subject spends too much time there, the rate of return falls below that provided by the VR. Melioration then says: "too much VI." Between the extremes falls the equilibrium point, where the alternatives provide equal rates of return per investment. This point conforms to the matching law, as do all equilibrium points produced by melioration.

To maximize reinforcement, the subject would have to find the highest point on the dashed curve. It is generally the case that at the maximum, the subject would be earning a higher rate of return from the VI than from the VR. To maximize, the subject therefore must resist its tendency to spend more time at a more lucrative alternative. Think of it this way: At the maximizing allocation, whatever would be gained by spending more time on the currently more lucrative variable interval is more than lost by the declining rate of return from that schedule. By now, several scores of experimental pigeons, rats, pied wagtails, and perhaps other species, have failed to maximize and have been drawn instead toward the matching point, which always means too much time spent on the VI.

No parametric, systematic study of human subjects earning reinforcement on the concurrent variable-interval, variable-ratio schedule has yet been published, partly because such experiments take months to complete. Anecdotal evidence suggests that the results might not differ fundamentally.

Suppose, for example, that you eat only caviar or hamburger for supper every night, that sometimes you eat one and sometimes the other, and that price is no object. If you chose caviar whenever you thought it would be even marginally better tasting, you would be meliorating and would be allocating your choices so that caviar tasted no better than hamburger on the average, and vice versa. Assuming that the pleasure of caviar declines more with consumption than that of hamburger, your aggregate pleasure could be far less than it is for those of us for whom caviar must remain a special treat. To get the maximum pleasure, we should husband caviar as a treat, rather than equate it to hamburger. If you ate more caviar than is optimal for maximum pleasure, you would be behaving like pigeons and rats on concurrent variable-interval, variable-ratio schedules, spending too much time on the variable interval.

More generally, most, if not all, of us imagine we would be happier wealthier than we are. We feel this way however wealthy we are. Are the very wealthy, then, very happy? Charles Murray (1988) concluded in his book that neither survey results nor common experience bears out the expectation of continuously increasing happiness with increasing wealth. Not only does more money not necessarily buy more happiness, it often seems to buy trouble. The "poor little rich girl," for example, does exist. Traditional morality counsels against the unfettered pursuit of material wealth but provides no scientific justification for its wise counsel. The melioration principle explains how having more money may lead to counterproductive equilibrium points; the maximization principle does not, at least not without quite a few epicycles.

Addictions and Other Pathological Choices

Figure 11.4 illustrates distributed choice, given just two alternatives. Visualizations of larger choice sets are harder to concoct, but the evidence suggests that matching and melioration apply as well to larger choice sets, albeit with greater formal complexity. However, for purposes of explication, we can always collapse larger choice sets to a choice between a particular alternative and "everything else." That is what Prelec and I have done in a theoretical discussion of addiction seen as a pathology of distributed choice (see Chapter 14).

Figure 11.5 shows how the reinforcement rates, here called "value," of two alternatives, 1 and 2, depend on allocation to alternative 1, where the total allocation, set at 1.0, is divided between the two alternatives. The addictive commodity, alternative 2, shown by the solid curve, declines in value per unit the larger the investment in it. The alternative competing with it, 1, shown by the dotted curve, rises in value the more it is chosen, up to some high level of consumption, beyond which it too falls with subsequent consumption. The joint returns from the two alternatives are plotted by the dashed curve.

Melioration predicts equilibrium at A, where choosing the addictive commodity, 2, predominates. A maximizer should choose the highest point on the joint function, B, but at B, the addictive commodity seems far more reinforcing than its competitors and is therefore, by melioration, likely to be chosen more. Exclusive choice of the nonaddictive alternative, 1, is labeled C.

This structure of reinforcement contingencies, Prelec and I suggest, portrays addiction inasmuch as the reinforcement returns at C are higher than those at A. In other words, in this environment, the subject benefits simply by being deprived of the option of choosing 2. Rational choice

A MODEL OF BEHAVIORAL ADDICTION

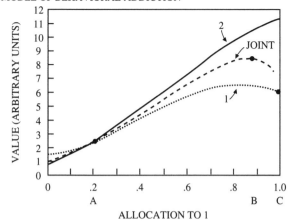

Figure 11.5 A model of behavioral addiction. *Note:* The average value of each of two competing commodities, labeled 1 and 2, declines as allocation shifts from 1 to 2. The joint returns from both commodities, the dashed curve, pass through their maximum at the filled point above B; equilibrium is that at A, and the filled point above C represents the joint returns when allocation is totally to 1. Adapted from Figure 14.4.

theory has trouble dealing with the idea that a person benefits by being deprived of an option, but common experience, and melioration, says it can happen easily.

Figure 11.5 shows addiction behaviorally, and that may be its special strength because addiction is fundamentally a pathology of behavior. An addictive commodity steeply loses average value per unit consumed at higher levels of allocation to it. At the same time, the alternatives to addiction also lose value when the addictive commodity predominates. As a result, the equilibrium point is at a low overall level of reinforcement, which is why addictions are considered counterproductive. The person may well know how counterproductive his or her way of life is but be helpless to escape from it.

A reinforcement structure containing the essentials of Figure 11.5 could arise in many ways. If the addictive commodity is a chemical, then excessive use may induce chemical changes in the brain, which in turn induce a tolerance that reduces the hedonic punch of a unit of the addictive chemical and also spoils the pleasures derived from other activities. But the commodity need not be a chemical. People who overwork may find that work is less fun than it used to be, but that it is still better than

not working. We may then say the person is a workaholic, implying irrational fixation on work. People sometimes find themselves trapped in
bad personal relationships: The keen pleasure in a romance may be long
gone, but alternative personal relationships have lapsed and are not easily reconstructed. We may not say that the person is addicted to his or
her partner, but we could.

Treating addiction often consists of a total prohibition against the
addictive commodity, a shift to C in Figure 11.5. As long as the person
never indulges in the addictive commodity, he or she will not experience
its superior returns in this region of allocations. Overall reinforcement
at C is suboptimal but superior to that at A. The optimal allocation, at
B, permits low levels of indulgence at which the addictive commodity
is still more pleasurable than its alternatives. Most people would find
B too hard to sustain, but that only shows they are meliorizers, not
maximizers.

Another approach to treatment is suggested by the analytic framework. A lowering of the value curve for the addictive commodity shifts
the equilibrium point toward the optimal allocation, as in Figure 11.6.

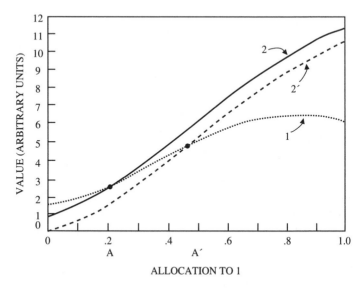

Figure 11.6 The equilibrium, at A, in Figure 11.5 shifts to A' when the value
of commodity 2 is reduced for all allocations. *Note:* The value of commodity 2 is
shown by the dashed curve. Consumption of 2 at equilibrium would be reduced by a
consumption tax on 2. (See also Figure 14.4.)

The dashed curve, labeled 2', illustrates how the addictive commodity curve would change if a use tax is applied to it. The new equilibrium, at A', is a shift away from excessive indulgence, toward optimality but not reaching it. A comparable improvement would result if the nonaddictive alternative earned a consumption bonus. Raising the height of the dotted curve likewise shifts equilibrium to the right. This tells us nothing more than we know from common sense. But conforming to common sense, with no loss of rigor, is one of the advantages of the melioration framework.

Conclusion

Rational choice theory and the melioration principle converge under certain conditions. For example, when the reinforcement rates associated with competing alternatives are independent of the frequency with which they are sampled, the subject maximizes by choosing the better alternative exclusively. We all choose $10 over $5. But melioration simply implies exclusive choice of the better alternative here. In the laboratory, this is approximated by concurrent ratio schedules; in economic theory, it is choice among constant probability alternatives, which is the classical economic paradigm. When human or animal subjects maximize in situations like this, as they often do, neither theory is strengthened at the expense of the other.

Rational choice theory adequately describes distributed choice in those situations in which the distributed nature of the choice is immaterial in the sense that the returns do not depend on the frequency of sampling. In many, though not all, other situations, it fares less well. Nothing in rational choice theory can tell us when it fares well and when it fails, but melioration does tell us. When melioration implies utility maximization (i.e., maximum reinforcement rate), rational choice theory adequately describes behavior.

But describing behavior does not seem to be the proper use for rational choice theory. Rational choice theory tells us how choice *should* be allocated, given a reinforcement or utility structure, not how it will be allocated. This normative function it serves admirably and usefully. A better analogy for rational choice theory than Newtonian physics is Boolean algebra.

George Boole, a 19th-century English mathematician, wrote his book (1854/1911) on the binary arithmetic named after him as a description of human reasoning: He titled it *An Investigation of the Laws of Thought.*

It turned out not to be much of a theory of human thought, but as a calculus of reasoning, the book was epochal. Instead of discovering the laws of thought, Boole had invented the algebra embodied in all modern digital computers, an algebra uniquely well suited to the behavior of a network of binary switching elements. One reason we find computers helpful is that our thought processes are often not Boolean. In just that way, we need rational choice theory because, as meliorizers, we often act suboptimally. How a meliorizer can make use of the guidance provided by rational choice theory is a complex matter, far from well worked out and beyond the scope of this essay (some sketchy notions are presented in Herrnstein, 1988, 1989).

Rational choice theory lies at the heart of not only modern microeconomic theory but also political doctrines that advocate minimal government—libertarianism and anarchism, for example. The idea is that, insofar as people behave rationally, they should be left to their own devices, except when collective behavior undermines individual interest, as when maximizing fishers overfish the waters or each individual decides that someone else should do a particular job, like serve in the army or build a road. But suppose people fundamentally and individually misbehave, as the evidence indicates they do. Then we would expect government to take account, not just of the defects of collective action, but of individual action as well, as David Hume (1777/1826) said more than 200 years ago. As old as it is, the idea remains unexplored and revolutionary, and it defines a conceptual frontier that students of the experimental analysis of behavior are uniquely well qualified to cross.

12

Behavior, Reinforcement, and Utility

> *Nature has placed mankind*
> *under the governance of two*
> *masters,* pain *and* pleasure.
> Jeremy Bentham, 1789

Jeremy Bentham, father of British utilitarianism, no more discovered "pain and pleasure" than Isaac Newton discovered that apples fall from trees. The simple facts of reward and punishment, like those of gravity, are part of ordinary experience. But, with the sentence quoted above, Bentham (1789) opened Chapter 1 of his *Introduction to the Principles of Morals and Legislation,* in which he gave such impetus to what he called the Principle of Utility that, two centuries later, it continues to be the flywheel of behavioral and social science, as well as of political, moral, and legal philosophy. Despite its age and its importance, questions remain unanswered about utility (or its synonyms in other disciplines, such as reward and punishment or reinforcement)— how to define it, to formalize it, to square it with ethics, to delimit its scope, to account for its evolutionary origins, to uncover its physiology, and to use it or not to use it for the good of oneself or the world.

Bentham and the other early utilitarians are, then, common intellectual ancestors to modern psychologists and economists, as well as to other students of human behavior. But, despite the common ancestry, economics and psychology soon diverged to such an extent that mu-

Originally published in *Psychological Science*, 1990, 4, 217–224. Copyright © 1990 American Psychological Society. Reprinted with the permission of Cambridge University Press. Versions of this essay have been presented at the Center for the Study of Public Choice at George Mason University, the Johnson Graduate School of Management at Cornell University, the conference on values sponsored by the Social/Behavioral Sciences Research Institute at the University of Arizona, and at the annual meetings of the Society for Mathematical Psychology. Thanks are owed to too many people to list by name, for providing the impetus for this work, for finding gaps in my argument, and for suggestions for improvement. There are doubtless gaps remaining, but fewer than there would have been without the help received. Thanks are, finally, owed especially to the Russell Sage Foundation, for providing support during 1988–89, and an environment rich in discussion on the general subject of behavioral economics.

tual incomprehensibility has become the major barrier to any attempt to unify the two disciplines at any level.

This essay will not try to review, let alone undo, two centuries of divergent intellectual evolution, much of which is just sensible division of labor among specialists. It focuses instead on the divergence at the level of theories of individual behavior. Here, the two traditions have arrived at conclusions that are often at odds. Both the economic and the psychological theories are incomplete, but at least one of them is fundamentally wrong, and is bound therefore to foster false conclusions. The conclusions affect not just how we think about ourselves, but also how we would design the environments that shape our behavior and that are, in turn, shaped by how we behave.

Although the principle of utility applies to individual behavior, economic thinkers naturally concentrate on exchange. Adam Smith's (1776) familiar example of a butcher selling meat to a customer exemplifies the molecule, if not the atom, of economic theory. Seller and buyer, both self-interested, establish the price at which each improves his lot by trading the money for the meat. This is the invisible hand at work, and what it was presumed to be doing is guiding the parties in exchange toward the best possible distribution of goods, services, and money. In time, economists elaborated this simple notion in so many distinctive ways that their approach to behavior is only barely recognizable to most psychologists.

Yet, it seems beyond question that the two disciplines must rest on a common foundation, namely a theory of individual human behavior. The following observations are offered in the hope of pinpointing where economics and psychology most relevantly diverge in their underlying theories of individual behavior.

By focusing on behavior in exchange, economics defined a subject matter that approaches individual behavior only indirectly, based as it is on multiple agents. The laws of individual behavior could be obscured when the data comprise exchanges rather than individual acts. For example, when the exchange takes place in competitive settings (a setting that includes, say, other butchers and other customers), competition per se may be decisive.

Economics has been formalistic or deductive when it characterizes individual behavior, rather than naturalistic or inductive. Economists have assumed that individuals are fundamentally rational, by which they mean that individuals tend to maximize utility. Theoretical economists invest time and effort in deducing the implications of utility maximization in myriad circumstances.

To outsiders, it is hard to tell whether economists believe in rationality as fact or definition, which is to say, by virtue of empirical evidence or as a useful tautology. Those may seem to be the only two alternatives, but an article in a leading journal of economics says, "Although the neoclassical hypothesis [i.e., utility maximization] is *not* a tautology, no criticism of that hypothesis will ever be successful" (Boland, 1981). Similarly, the prominent economist Gary Becker, wrote, "The combined assumptions of maximizing behavior, market equilibrium, and stable preferences, used relentlessly and unflinchingly, form the heart of the economic method" (Becker, 1976, p. 4). Those sentences capture well the hold of the maximization assumption over economic theory.

Economists recognize that the rationality assumption is often contradicted by actual behavior, but rarely does this inspire a search for some other fundamental assumption about behavior (for discussions of this mismatch between theory and data from the psychological and economic viewpoints, respectively, see Tversky and Kahneman, 1986; Zeckhauser, 1986). Rather, to the extent that the irrationalities cannot be blamed on sheer random error, they typically lead to a search for subsidiary principles, concerning systematic distortions of perception of probability and time, failures in memory, insufficient information, erroneous rules of thumb for calculating outcomes, and the like.

In short, much seems to be at stake for economic theorists in their assumption of utility maximization. Possibly, what is at stake is a further, rarely stated assumption—that any assumption about behavior other than that of maximization would exact too great a price in lost rigor, as if the clarity and precision of formal economic analysis under the assumption of rationality is worth a certain amount of error in the prediction of behavior.

The deductive economic approach has exploited the implications of a concept of an economic equilibrium. The idea that a system is drawn toward certain well-articulated outcomes defines mathematical structures of proven worth in the physical sciences, where they have given shape to such concepts as entropy and least effort. But the idea of an equilibrium state for the behavior of an individual actor says little about how an actual person comes to behave in a given way. Economic theory is not committed to any particular process or mechanism guiding the flow of behavior, which gives rise to the list of the characteristics I wish to note.

Somehow we maximize, the theory says, and how we do so is not the concern of economic theorists. The conventions of economic theorizing require only that the theorist show that someone's investments or behavior do or do not satisfy the assumption of optimality. Behavior is seen to be unaccounted for until it can be shown to be consistent with utility maximization or it is explained why not.

We can contrast this list with the approach of behavioral psychology, from which the competing theory arises. Behavioral psychology has searched for the processes that control behavior, rather than for the equilibria that those processes might produce. Lacking a presupposition about the ends of behavior, psychology has been far more inductive than deductive.

The occasion for this essay is a body of evidence from the psychological laboratory consistent with the notion that an individual's allocation of behavior tends to approach an equilibrium, but not necessarily the equilibrium required by utility maximization. Many of the systematic departures from rationality that are part of everyday experience can be interpreted as exemplifying a principle of equilibrium that differs from utility maximization. The tools of formal analysis are as readily applicable to this new principle of equilibrium as the rationality assumption of standard economics.

Melioration, Matching, and Suboptimality

The present approach takes the view that behavior is generically suboptimal, though still orderly, and that optimality is the exception rather than the rule (see Chapters 10, 14; Prelec, 1982). The governing theory for this approach is based on numerous laboratory experiments exploring the law of effect in animals and humans. The theory is built around the idea that choice is guided by the law of effect to produce a particular sort of limited optimization, one that happens to be susceptible to serious lapses from rational choice.

Data from the behavior laboratory crystallize around what has been called the *matching law* (see Chapter 1; Herrnstein, 1970), which has been observed in several hundred experiments on various species, including human (see Chapter 2; Bradshaw and Szabadi, 1988; Davison and McCarthy, 1988; de Villiers, 1977; Williams, 1988). The matching law states that, at equilibrium, an individual's behavior is distributed over alternatives in the choice set so as to equalize the reinforcement returns per unit of behavior invested, measured in time, effort, or any other dimension of behavior constrained to a finite total. Any systematic deviation from equality in reinforcement per unit of behavior invested is destabilizing, driving behavior toward an equilibrium in which the deviation is absent.

The dynamic process that yields matching at equilibrium has been called *melioration* (see Chapters 4, 14; Herrnstein and Vaughan, 1980; Prelec, 1982). The notion of melioration is simply a restatement of the principle of reinforcement itself: other things equal, more reinforced responses occur more often than less reinforced responses. A rise or a fall in the reinforcement of a response causes the rate of occurrence of the response to change in the same direction. Should there be an inequality in unit returns from two alternatives, behavior "meliorates," redistributing itself toward the more lucrative, hence stronger, alternative. The melioration process continues until the stronger response displaces all others, or, because the reinforcement returns from an alternative may depend on its level of occurrence, equilibrium is attained with several alternatives left in the response set, each yielding the same returns per unit at a given allocation among them, which is the matching law. The most salient observable difference between the matching law and the equilibria implied by standard economic analysis is that, at a matching equilibrium, a person may not be allocating behavior optimally, given his own utility functions, as will be shown below.

As noted earlier, standard economic analysis and reinforcement theory may both properly claim descent from the utilitarian conceptions of the 18th and early 19th centuries. Both of them depict behavior as being adaptive. They differ in that the adaptation implied by the matching law is limited by the person's tendency to credit, as it were, the returns he receives to a particular response alternative, rather than to keep a global account of returns across his entire repertoire, which the modern theory of rational choice requires him to be able to do. This limitation in mental bookkeeping entails a limitation in our general capacity to discover optimal allocations of our behavior, although particular circumstances determine whether the limitation is grave, trivial, or, in certain cases, nonexistent.

It should soon be evident that the fundamental difference between matching and utility maximization is that matching is based on average returns (in utility or reinforcement) over some extended period of activity, while maximization requires a sensitivity to marginal returns at each

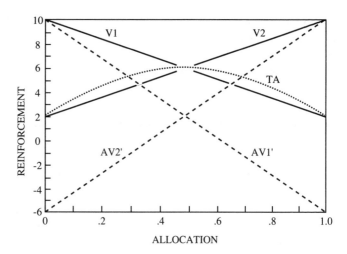

Figure 12.1   Showing the interaction between several reinforcement variables and the allocation to one or the other of two mutually exclusive, exhaustive alternatives, in a hypothetical procedure. The abscissa gives the allocation to the 1-alternative; the 2-alternative is its complement. $V1$ and $V2$ plot the rates of reinforcement per unit of allocation to the 1-alternative and the 2-alternative, respectively. Their intersection is the matching point. $AV1'$ and $AV2'$ plot the marginal changes in the number of reinforcements associated with a marginal change in allocation to the 1- and 2-alternatives, respectively. Their intersection is the optimal allocation, also shown by the maximum value of TA, which plots the total number of reinforcements obtained from both alternatives. Equations are given in the chapter appendix.

moment.[1] Where the marginal and average returns to response alternatives are equal or bear certain other specifiable relations to each other, we would expect to find no large difference in the predictions of theories relying on one or the other of them (see Chapter 14, for the general case). The customary economic assumption about the marginal utility of a commodity (or other sources of utility) is that it decreases as the rate of consumption increases. Figure 12.1 illustrates allocation across two

---

1. For the distinction between rates of reinforcement on the average and at the margin, consider a subject distributing its behavior across a set of alternatives. The average rate of reinforcement for each alternative is the ratio between the total amount of reinforcement received from the alternative and the time invested in the alternative, at that allocation of behavior. The rate of reinforcement at the margin for each alternative, at a given allocation of behavior, is estimated by the change in reinforcement received from that alternative associated with a marginal change in the time allocated to that alternative, which is to say, the derivative of the function relating the reinforcements received from each alternative to the time spent engaging in it.

such commodities in the kind of chart that matching theory usually focuses on.

It is assumed for this example that the subject is constrained so that consumption of the two commodities, 1 and 2, is exhaustive and mutually exclusive, and that the average utility (or value of reinforcement) obtained from each commodity during an extended period of observation is inversely and linearly related to how much of it is consumed in the period. The lines designated as $V1$ and $V2$ in Figure 12.1 show the reinforcement received from one or the other of two commodities per unit of allocation of behavior invested in it, given a long-term allocation in commodity 1 represented along the x-axis (the consumption of commodity 2 being the complement). With a linear decrease in the average yield as a function of allocation to an alternative, the marginal returns also decrease linearly, albeit twice as rapidly (these relations are explicated in the appendix). The marginals at each long-term level of allocation are traced by the lines labelled $AV1'$ and $AV2'$.

Diminishing returns may arise externally or internally. The reinforcement per unit invested may deplete with the average time spent consuming it or with the rate of consumption, over a period of observation. In repetitive choices between, for example, two berry bushes over some period of time, the average quality and number of berries per visit may bear an inverse relation to the average rate at which we visit a bush. But reinforcement returns may diminish, or otherwise change, because of internal processes—pizza too many evenings or too many trips to a museum may take the edge off the pleasure. Repeated exposures may also enhance reinforcement: we may come to love Mozart more if we listen to it more, over some range of consumption rates. Any of these cases may be represented as in Figure 12.1, with appropriately drawn functions.

The curve labelled TA in Figure 12.1 depicts the aggregate reinforcement returns over the two commodities. Standard economic analysis, as well as common sense, says we should find the maximum on this curve, at the point where the joint returns are best and the marginal returns are equal. In this environment, melioration implies the same outcome, for the maximum of the aggregate curve coincides with the crossing $V1$ and $V2$ curves, which is the equilibrium point conforming to the matching law.

The picture in Figure 12.1 needs only a slight change in order to highlight the fundamental difference between matching and maximization. In Figure 12.2, the two commodities still yield linearly diminishing average returns with increased consumption, but because of their mutual asym-

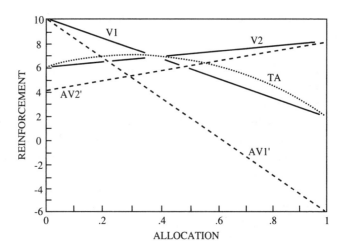

Figure 12.2   Except for differences in the equations plotted, the same as Figure 12.1. See text and appendix for details.

metry, the maximum of the aggregate returns (the maximum of the TA curve or the intersection of the $AV1'$ and $AV2'$ lines) has shifted toward alternative 2, relative to the matching point at the intersection of the $V1$ and $V2$ lines. Common sense is equivocal here (shall we equalize the averages or the marginals?), but the evidence from the laboratory suggests that in environments like this, no less than that depicted in Figure 12.1, subjects equalize the averages and thereby match, but here they do so at some cost in overall gains.

Note, further, that in both figures the aggregate returns, the TA curves, are relatively flat in the intermediate range of allocations. Flatness means that overall reinforcement is insensitive to the overall allocation of behavior, while the difference in average returns for each commodity, the $V1$ and $V2$ lines taken separately, interacts sharply with the allocation. The behavior of subjects in controlled experiments is usually narrowly concentrated around the matching point (Williams, 1988), even when the overall returns are not much affected by allocation. Indeed, even when the aggregate returns have zero slope in the region of observed allocations, subjects still approximate matching (Vaughan and Miller, 1984).

The laboratory evidence, then, supports three related propositions. First of all, subjects allocate their behavior optimally when doing so also satisfies melioration and the matching law. Second, when matching

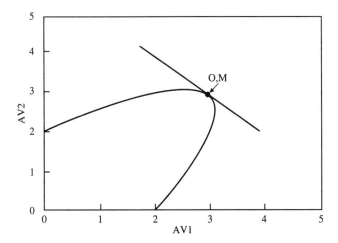

Figure 12.3   The curve shows how the reinforcements received from the two alternatives trade off with each other, given the procedural constraints implied by the functions plotted in Figure 12.1. The linearly decreasing line is the locus of equal total reinforcements (i.e., the assumed indifference curve) tangent to the constraint curve, namely, a total of 6 reinforcements. Filled circle is at the optimal (O), as well as the matching (M), distribution of reinforcements, given the constraints.

and maximization make divergent predictions, behavior is often closer to matching than to maximization. Finally, in situations in which maximization does not narrowly constrain allocation, subjects still conform relatively sharply to the matching law (see Chapter 10).[2]

Figures 12.3 and 12.4 repeat the examples depicted in Figures 12.1 and 12.2, respectively, using coordinates more familiar to economists and to those psychologists who have lately adopted economic approaches to behavioral data. Since the two alternatives in each of those examples provide access to the same reinforcement, the reinforcements from the alternatives are assumed to be perfectly substitutable, and the indifference curve is therefore linearly decreasing and complementary

2. This, of course, is the author's reading of the evidence. Others may, and do, differ (see Commons, Herrnstein, and Rachlin, 1982, for some competing theories among behavioral psychologists). The author's reading appears, at least for the moment and in regard to the matching law itself (if not to the melioration principle), to approximate the predominant one (e.g., Davison and McCarthy, 1988; Williams, 1988), but it is by no means unanimously accepted. It is also a reading based primarily on the results of experiments using animal subjects. The laboratory data from human subjects are generally supportive, but would not on their own substantiate matching as the principal challenge to utility maximization.

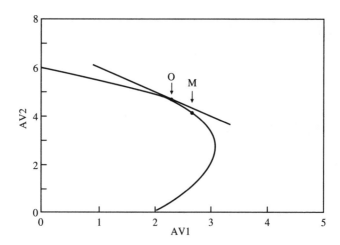

Figure 12.4   Same as Figure 12.3, for the functions plotted in Figure 12.2. The assumed indifference curve is at a total of 6.9 reinforcements. Filled circles at O and M show the optimal and matching distributions, respectively.

(see Appendix). Figure 12.3 shows how the linearly decreasing utility in-difference (i.e., equal reinforcement) curve implied by Figure 12.1 forms a tangent with the "budget constraint" in this procedure, which is to say, the constraint imposed by an experimental session in which the subject has a given amount of time or behavior to allocate to one or another of two alternatives. The subject earns reinforcement from both alternatives, with the number depending on how behavior is allocated and on the $V1$ and $V2$ functions plotted in the previous figures. The filled circle plots the allocation that conforms to matching (i.e., M) and to the utility (or reinforcement) maximum (i.e., O).

Figure 12.4, however, exemplifies the challenge to maximization the-ory implicit in matching. The reward structure in Figure 12.2 also im-plies linearly decreasing indifference curves, but now the tangent to the budget constraint (that is, the optimal allocation, O) is no longer at the matching point (M). The evidence suggests that, in situations like this, subjects allocate suboptimally, approximately matching. The separation between matching and maximization can be large, small, or nil, depend-ing upon the contingencies of reinforcement for the alternatives. The location of the matching point is not determined, nor even identifiable, by the variables in standard indifference charts like those in Figures 12.3 and 12.4.

If utility maximization is generically violated by behavior, it is natural to ask how so momentous a fact of nature could have been overlooked by the proponents of rational choice theory as it is applied in economics. The answer is not that economists have generally carved off for themselves the rational bit of human behavior, leaving the irrational residue for the other social sciences to grapple with. As Hirshleifer has written, explaining the "expanding domain" of economics, *"There is only one social science . . .* Thus economics really does constitute the universal grammar of social science" (Hirshleifer, 1985, p. 53, emphasis in original).

One must agree with Hirshleifer that the logic of modern economics makes it inherently imperialistic: "It is ultimately impossible to carve off a distinct territory for economics, bordering upon but separated from other social disciplines. Economics interpenetrates them all, and is reciprocally penetrated by them" (Hirshleifer, 1985, p. 53). It is, in fact, the underlying assumption of rationality, the very assumption I suggest is false, that ties the social sciences together, in the view of modern economic theory.

Utility Is Invisible but Essential

The Principle of Utility can still be a matter of debate 200 years after Bentham's enunciation of it because utility is invisible. We may observe behavior and its objective consequences, but, however carefully we observe or deeply we probe, utility itself will remain out of reach of direct observation. The invisibility of utility is a particular instance of the inscrutability of subjective states in general. Like force in physics, utility means a relationship between observables, but is not itself observable.

We have strong intuitions about utility. We expect people to place positive subjective value on food, companionship, the admiration of others, and negative subjective value on losses of status, wealth, health, and so on. Most of the time, most of us may behave accordingly, hence our intuitions. But a person may differ in these respects from other people, at least sometimes. From time to time, we surprise each other or ourselves by our choices. When we defy intuition, it is impossible, on that one occasion, to tell whether the source of the surprise is an uncommon, aberrant, even potentially self-destructive, utility function, or a generic principle of choice that occasionally guides behavior in some way other than rationally, given the person's own utility function. It is only by looking at behavior in many situations, varying along many dimensions, that we are able to build a consistent case for one or the other of those explanations.

For the traditional economist, committed to utility maximization as the fundamental choice rule, explaining behavior has had only one degree of freedom, the utility functions themselves. Only postulations about the nature of the motivational substratum are at hand for saving the theory (see, for example, the treatment of addiction by Becker and Murphy, 1988). Many economists have lately been willing, even eager, to expand the conception of utility beyond the limits of mere wealth, to the "deeper successes that are not made of gold and silk," in the words of George Bush's inaugural address (quoted in the *New York Times* by Passell, 1989, in a quite pertinent article on Robert Frank's book, *Passions within Reason*, 1988), or in such works as Scitovsky's (1975) book on *The Joyless Economy* and many others. The arguments about a surprising bit of behavior are often concerned with competing inferences about the underlying utilities (see, for example, Caporael, Dawes, Orbell, and van de Kragt, and associated commentary, 1989, in regard to "selfishness"). But few economists have been willing to explore the possibility that the explanation lies elsewhere than in the utility functions.

For behavioral psychologists, in contrast, there have always been two degrees of freedom, for the behavioral rule itself was in question, not just the motivational substratum. Only by systematically varying the circumstances in which choices are made—by doing controlled experiments, in other words—can we find a rule for interpreting behavior that will explain behavior in terms of its consequences, whether or not the behavior violates our intuitions.

Conclusion

The empirical tradition of behavioral psychology gives us a picture of choice that is partially, though not wholly, inconsistent with the rational choice theory of modern economics. The similarities are worth noting. In both theories, behavior is governed by its consequences and the relevant consequences can only be inferred from behavior. Neither discipline believes it can get at the essential but subjective hedonic facts, the utility functions or the drives and the associated rewards and punishers, except by how they are revealed in behavior.

The bootstrap operation differs, however, in the two disciplines. Economics assumes generic rationality. Behavioral psychology assumes nothing generically about behavioral equilibria, but, in the course of experiments on human and animal behavior, it discovered the law of effect, which was then refined into the matching law.

We have suggested that matching is the equilibrium point produced by melioration, a process in which response classes compete for the organism's behavioral investment, coming into equilibria where the benefits per unit investment are stably equalized (see Chapter 14 for stability conditions). The matching law is as capable of being the basis for inferences about utility as the principle of utility maximization, but the inferences may differ. The matching point may be suboptimal, but it need not be. The class of maximizing equilibria can be shown to be a subset of the class of matching equilibria, given the process of melioration (see Chapter 14).

In traditional economics, the utility functions are inferred but not explained. Economics provides what Hirshleifer called a "grammar" for deriving and expressing the utility functions, but no economic theory says where they come from. The evolutionary origins and physiological mediation of the rewards and punishers for species or individuals are, in contrast, traditional matters of concern for behavioral psychologists. And, in addition to those biological roots, behavioral psychologists study the modification of the rewarding and punishing powers of stimuli by Pavlovian conditioning. For motivation, as for other elements in their theories of individual behavior, psychology has less grammar and more laboratory data than economics.

A fundamental difference between the disciplines is the status of response classes. In a utility-maximizing theory, no fundamental significance attaches to how the organism subdivides its own behavior, except insofar as one or another "accounting scheme" is more or less inherently costly or pleasurable. The theoretical maximizer operates by the principle of "universal accounting" (Luce, 1990), treating costs and benefits across all alternatives as fungible. According to the principle of melioration, however, the accounting scheme is decisive in determining the equilibrium point, as described above (and discussed at further length in Chapter 14).[3] No comprehensive theory of how organisms arrive at particular accounting schemes is at hand, a gap in theory that will in time need to be filled.

The interaction between choices and returns, we also noted, may arise in the environment or in motivational states. It is plausible that the environmental sources are public and visible: we can all see that too many

---

3. The concept of "framing," well known to economists but uncovered by psychological experiments on behavior (e.g., Tversky and Kahneman, 1981, 1986), is another manifestation of the significance of the way behavior is being partitioned.

visits to the fishing grounds deplete the stock, hence reduce the average reward per visit at a later time.

For intra-individual interactions grounded in motivational states, it may be less possible to accumulate comparable public wisdom or, at any rate, to supply it in a timely fashion. We may know as a general rule that eating or drinking too much or exercising too little can be dangerous, but suboptimal rates of indulgence may show no visible traces until considerable utility has been forgone. The suboptimalities arising in motivational externalities (the economist's "changes in tastes") are therefore likely to remain more uncontrolled, hence of greater potential danger, than those in environmental externalities, at the individual level.

Because of the invisibility of utility, economists have argued that the ostensible irrationalities involved in obesity, addiction, etc., are really not irrationalities at all, but only the result of peculiarities of one sort or another in utility functions. Economic Man may harm himself objectively, but he cannot fail to satisfy the demands of utility maximization, according to modern economic theory. While not excluding the possibility of peculiar and self-destructive utility functions, the theoretical alternative proposed here holds that the principle of action itself, above and beyond the sources of utility, can lead to maladaptive behavior. The present alternative says that people may need help in their pursuit of subjective satisfaction, let alone objective well-being.

In short, after 200 years, we descendants of Bentham can agree about the primacy of pains and pleasures; we can agree even that, as Hirshleifer more recently said, there is only one social science. But now, to pursue that social science, and to use it to design social institutions, we need to reconcile the divergent answers provided by empirical approaches, such as that of behavioral psychology, with the formal structures of economic theory.

Appendix

Equations for the variables plotted in Figures 12.1–12.4 are given below. In these equations, allocation to the 1-alternative is represented by X, and to the 2-alternative, by 1-X.

$$V1 = b - a(X),$$

$$V2 = b' - a'(1 - X),$$

$$AV1 = X(V1),$$

$$AV2 = (1 - X)(V2),$$

$$AV1' = d(AV1)/dX,$$

$$AV2' = d(AV2)/d(1 - X),$$

$$TA = AV1 + AV2.$$

For Figures 12.1 and 12.3, the parameters are: $a = a' = 8$; $b = b' = 10$. For Figures 12.2 and 12.4, the parameters are: $a = 8$; $a' = 2$; $b = 10$; $b' = 8$.

$V1$ and $V2$ give the average rates of reinforcement per unit of allocation to each of the two alternatives. $AV1$ and $AV2$ give the numbers of reinforcements from each alternative at each allocation. $AV1'$ and $AV2'$ give the marginal change in reinforcements for each alternative given a change in allocation to it. TA gives the total number of reinforcements from both alternatives, at each allocation.

The indifference curve in Figure 12.3 plots a segment of the line:

$$6 = AV1 + AV2.$$

The indifference curve in Figure 12.4 plots a segment of the line:

$$6.9 = AV1 + AV2.$$

13 | Experiments on Stable Suboptimality in Individual Behavior

In several recent experiments,[1] subjects made choices in a way that supports the idea that choice is governed, either sometimes or always, by a principle that does not necessarily maximize utility, as the subjects themselves would reckon their utility. In other words, they behaved irrationally, and their irrationalities seemed to be systematic and motivated, not just a matter of behaving carelessly or erroneously.

Representative Data

The subjects in the experiments worked for money by repeatedly choosing one of two alternatives in procedures in which the long-run pattern of choice interacted with the rate of payment for each alternative. The subjects earned less money than they could have, and they did so when there was no plausible compensating nonmonetary gain. Their suboptimalities were often consistent with a principle of choice well established by hundreds of experiments mainly on animal behavior (Davison and McCarthy, 1988; Williams, 1988; for related experiments on human subjects, see Bradshaw and Szabadi, 1988).

Subjects (mostly students) were recruited at the University of Chicago and Harvard University with posters announcing the possibility of earning a few dollars in a brief session studying decision making. A subject

Originally published in *American Economic Review*, May 1991, *81*, 360–364. All of the experiments to be described grew out of an experimental design first used in the cited work, but much extended in discussions at the Russell Sage Foundation during 1988–89, with Ronald Heiner, George F. Loewenstein, Dražen Prelec, Howard Rachlin, and William Vaughan, Jr. The research was itself supported by the Russell Sage Foundation. The collaborators for particular procedures are noted in turn.

1. See Herrnstein, Prelec, and Vaughan (1986).

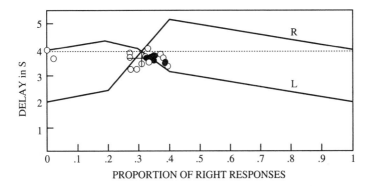

Figure 13.1   Experiment 1.

sat in front of a computer monitor in a small, quiet chamber. The instructions explained that striking the left or right arrow on the keyboard would release a coin from one or the other of the two boxes portrayed on the screen. Each subject made about 300–400 choices of the left or right key for money, after about 100–200 practice choices under the same conditions. Striking the left or right arrow key always earned a coin worth one cent each with a probability of 1.0. The subject was told that payment would be equal to the total value of the coins earned, and also that it was a good idea to try to earn coins rapidly. From time to time, a message would appear on the monitor telling the subject how much time was left until the end of the session.

Figure 13.1 shows the contingencies of reward in experiment 1.[2] It shows the time taken for a coin to fall from the right and left boxes (R and L, respectively), as a function of preceding 40 choices that were of the right key. The dotted line is the average time weighted by the abscissa value. The open circles plot individual subjects and the right angle cross shows the average subject. On the vertical axis is the time (in seconds) for a coin to drop into reservoirs from either the left or right box. While one coin is dropping, no new coin could be earned. A session comprised $6\frac{2}{3}$ minutes of practice followed by 15 minutes earning money to keep.[3] Coins fell at a speed that depended on the subject's recent choices.

---

2. This experiment was done by Vaughan and Herrnstein.
3. These times refer to elapsed times while coins were dropping, not continuous clock time.

The abscissa value is the proportion of right key choices in the sub-ject's preceding 40 choices, hence updated after every choice. The time it took for a coin to fall depended on where along the x-axis the choice was made. Suppose the preceding 40 choices had been evenly divided be-tween left and right keys and the subject chose the right key. It would take the coin almost 5 seconds to drop. A coin from the left box would have fallen in almost 3 seconds. Indeed, so long as the proportion of right choices in the preceding 40 trials is above .4, coins from the right box take 2 more seconds to fall than coins from the left. Below a proportion of .2, coins from the left box take 2 more seconds to fall than coins from the right. At a proportion of .3 choices of the right key, coins from either box take 4 seconds to fall.

The reward contingencies were chosen so that the weighted average time to drop, traced by the dotted line, was 4 seconds, no matter what the subject did. Consider the two extremes and the midpoint. At the midpoint, half the coins were taking about 5 seconds and half taking about 3 for an average of 4. At the extreme right, all the coins were falling from the right box and taking 4 seconds, and inversely at the left extreme, where all the coins were falling from the left box. For every point along the x-axis, the weighted average was 4 seconds. Given an averaging window of 40 trials, and an average of 4 seconds per trial, the subject could experience the entire reward functions in 160 seconds of coin-dropping time.

What is the rational strategy here? Indifference, which is to say, a 50–50 split, seems indicated, since all strategies earn equal amounts of money. Instead, of the 24 subjects in Experiment 1 and displayed in Figure 13.1, 22 came within a few percentage points of equalizing the falling speed of right and left coins, which is at .3 on the x-axis.[4] Each point shows the average allocation of a single subject during the latter half of the choices made by the subject during the one session he or she ran. The average of all subjects, shown by the crossed lines, was a 30.08 percent choice of the right.

Although the strategy is not rational in any obvious sense, it makes in-tuitive sense. Picture yourself in the situation and imagine that you find coins from the right box falling more rapidly than coins from the left. Consequently, you start choosing the right side more often. Gradually,

---

4. Points deviate from the theoretical average line because, with the averaging proce-dure we used, the current ordinate value for a given proportion of right and left choices depends slightly on the precise sequence of choices within the previous 40 trials.

coins from the right slow down and coins from the left speed up, and you find that those from the left are now falling more rapidly than those from the right. Your choices will now begin to swing back toward the left, since those coins are falling more rapidly. This strategy leads to an oscillation around 30 percent choice of the right key, and, judging from the data, it is the one adopted by virtually all our subjects. At 30 percent, the subject has equalized the average rate of returns from the two choices, namely a penny per 4 seconds. The equalizing of average rates of return from competing alternatives was first observed experimentally in 1961 in an experiment on pigeons earning bits of food (see Chapter 1). It has since been approximately confirmed in several hundred experiments on various species and is called the matching law because it calls for a match between the ratio of behavioral investments to the yield of those investments across all competing alternatives.

The process of comparing the rates of return and shifting toward the alternative that is currently yielding the better return is called melioration (see Chapters 4, 14; Herrnstein and Vaughan, 1980). Melioration has the inevitable effect of stabilizing in the vicinity of the matching law, and it also seems to be the process that most people spontaneously invoke when they are asked to describe their own choices.

Here is a second experiment.[5] Again, volunteer subjects earn coins by striking left or right keys, the coins are worth one cent each, and the time to drop determines how much is earned in a 15-minute session. And again, the time to drop depends on the distribution of choices in the preceding 40 trials, as shown in Figure 13.2. The x-axis expresses the subject's previous 40 choices as a proportion of choices to the right key. The three lines of Figure 13.1 are perfectly superimposed in Figure 13.2. Whatever a subject has done in the preceding 40 choices, coins fall at the same speed from either box.

Suppose, for example, that a particular subject has divided his choices 50–50 and now makes a choice. Whether he chooses left or right, the coin will take about 3.6 seconds to fall. If he had been choosing only the left for the previous 40 trials, a coin from either the left or right would take 5 seconds to fall. The fastest possible speed is 3 seconds, which is what either the left or right would take if 30 percent of the preceding 40 choices had been of the right. Since the delays are always equal for right and left coins, the weighted average delay is also equal to them.

5. Also by Vaughan and Herrnstein.

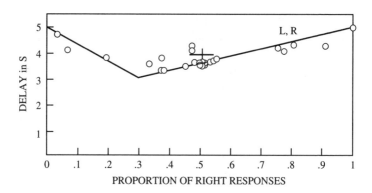

Figure 13.2   Experiment 2.

The most profitable strategy is at the lowest point on the line, where coins are all taking 3 seconds to fall. But instead of choosing right 30 percent, the 24 subjects are spread out broadly, approximating a 50–50 division between left and right. The average choice, the crossed line, was 50.08 percent on the right. Many of the individual subjects are also near indifference, as well as their average. Indifference is what the process of melioration implies, for the average yield for left choices is always the same as that for right choices.

Melioration, rather than maximization, appears to have driven the subject's choices in these two experiments. In Experiment 1, melioration defined a particular allocation of choices, that subjects approximated, even though all allocations yielded equal overall earnings. In Experiment 2, maximization called for a particular allocation, but melioration did not; the subjects seemed to be indifferent to the effect of their allocations on earnings. The coin-dropping times involved in the two experiments were essentially the same. The sessions lasted equal times and the amounts of money were the same. In Experiment 1, a 2-second difference was enough to determine behavior; in Experiment 2, a 2-second difference was ignored. From these experiments, we could postulate that a 2-second difference controls behavior if, and only if, it is relevant to melioration.

But neither experiment pitted melioration against maximization in a way that would decisively affect earnings. In the first experiment, melioration caused the subject to obey the matching law when doing so had no effect on money. In the second, indifference to the two alternatives cost the subjects something like a half-second per choice, which may be

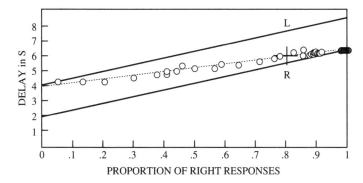

Figure 13.3   Experiment 3.

considered negligible. Let us now consider a third experiment in which melioration implies the maximally *inefficient* allocation of choices.

The procedure is summarized in Figure 13.3.[6] As before, the time taken for a coin to drop depends on the subject's choices. The abscissa here is the subject's allocation to the right key during the just-preceding ten choices. The entire reward functions could be experienced in a minute or less. Coins were again worth one cent. Coins from the right box always took 2 seconds less to fall than coins from the left. However, the more choices there were of the right box in the preceding ten trials, the more slowly coins dropped from either box. The optimal strategy would have been to choose the left key on every trial (disregarding the negligible "end effect," which no subject exploited). Melioration dictates choosing right on every trial: it dictates maximal inefficiency. For the 56 subjects, the average chose the right on 82 percent of the trials. A large majority of subjects "learned" the least efficient strategy. At most, one subject discovered the maximizing strategy. Obeying the principle of melioration gets relatively expensive here.

Conclusions and Implications

In some of the procedures we have examined, the average performance is closer to maximizing earnings than here presented. For example, when the interaction between choice and consequence is more rapid (when the averaging window is, say, 3 or 6 preceding trials rather than 10 or 40),

6. Experiment by Loewenstein, Prelec, Vaughan, and Herrnstein.

performance approximates maximization. This suggests an exceedingly limited capacity for taking account of the effect of one's own choices on marginal rates of return. In contrast, our subjects were quite sensitive to average rates of return, as shown by the close approximation to the matching point in Experiment 1.

Another factor appears to be the dimension of reward. Rate of money, in the guise of a falling coin on a computer monitor, was the relevant dimension here. In other experiments, coin denomination has been the relevant dimension. When the preceding choices determined what the coins were worth, performances were closer to maximization. It takes a larger averaging window to produce a given degree of suboptimality with coin denomination as the reward dimension than coin rate.

Yet another class of factors affecting performance concerns the degree to which the alternatives are presented as discretely competing responses. The left and right arrow keys were perfectly correlated with coins dropping from left and right boxes, respectively. But when subjects were able to choose a certain proportion of coins from one box or the other, much like a subscription to left or right coins over some interval of time, performance shifted toward maximization. It seems that anything that makes it easier for a subject to redefine the response categories will make it easier to maximize.

For some experimental conditions, subjects span the range between maximization and matching. Individual differences have occasionally looked trimodal—with some subjects near the matching point, some near the maximizing point, and some near 50–50, perhaps indifferent or confused (or both). Intersubject variability seems to peak under conditions that are transitional between those producing maximization and those producing melioration. For example, as the averaging window is increased from, say, 6 to 40, the intersubject variability rises, then falls, as the average subject moves from maximizing earnings to obeying the matching law.

At the beginning of this essay, I suggested that the evidence points toward generically suboptimal human choice under at least some conditions, and possibly under all conditions. A word of explanation about the latter possibility seems in order, inasmuch as the evidence seems only to say that people are sometimes irrational in their choices, not that they are always irrational. But there is a difference between generic suboptimality and universal irrationality. It should be obvious that, for any situation for which a maximization strategy exists, there also exists a definition of response alternatives such that the process of melioration leads to maximization. Given this, it could be argued that people are, in fact, always following the principle of melioration and that, when they are being rational, they are being rational only incidentally. This argument would be worth making only if there were also a well-developed theory of how a subject defines response alternatives, and we are not yet at that point.

Indeed, we do not know the learning algorithm that produces matching. Melioration implies a comparison of current average yields from competing alternatives, but beyond that vague characterization, the process remains to be explicated. At present, the prospects for doing so are unclear. We have examined the trial-by-trial choice patterns of our subjects and the only conclusion I can draw is that people appear to have many different ways to reach the matching point. Some subjects converge rapidly on an allocation, others scan the range of allocations more or less systematically, and still others do not behave in a way that allows simple characterization. For animal subjects as well, individual variation in the trial-by-trial choices is much greater than the variation around the matching point (see Bailey and Mazur, 1990). In the behavioral, as in the physical, sciences, the stable equilibria are often discovered before the dynamic processes that produce them.

14 | Melioration: A Theory of Distributed Choice

RICHARD J. HERRNSTEIN AND DRAŽEN PRELEC

In modern expositions of the theory of rational choice, the notion of what constitutes an object of choice is kept deliberately vague, so as not to restrict unduly the range of possible problems to which the theory will be applied later on. A recent advanced textbook on consumer behavior (Deaton and Muellbauer, 1980), for example, initially defines the objects of choice as "individual purchases of commodities" (p. 269), but this concrete definition yields quickly to the more abstract notion of a commodity "bundle," which is rarely obtained through an individual purchase. Implicit in this exchangeability of terms is the supposition that in applying the theory it does not matter whether: (A) the "choice" corresponds to an actual decision, made at a specific point in time; or (B) the "choice" is an aggregate of many smaller decisions, distributed over a period of time.

In this paper, we develop a theory of individual choice called *melioration*, for which the distinction between choices of type A and type B is critical, and which implies that choices of type B may be reliably and predictably suboptimal, in terms of the person's own preferences. If true, this would imply that preferences as revealed in the marketplace may be a distortion of the true underlying preferences whenever the measured

Originally published in *Journal of Economic Perspectives*, Summer 1991, 5, 137–156. The authors are grateful to Jerry Green, Ronald Heiner, Vijay Krishna, George Loewenstein, Howard Rachlin, Richard Thaler, William Vaughan, Jr., and Richard Zeckhauser for valuable discussion of the ideas presented here. The concept of melioration itself was initially formulated in collaboration with Vaughan (Herrnstein and Vaughan, 1980), and some of the most striking experimental substantiations of it are in his experiments (Vaughan, 1981). We owe thanks, too, to the Russell Sage Foundation for providing us with support and an environment conducive to developing these ideas.

economic variables are aggregates of a stream of smaller decisions; the extent and direction of the distortion is then something that the theory will need to explain.

Our theory draws support from two sources. First, experimental psychologists have accumulated much data about repeated choice over the last 25 years, indicating that for both human and animal subjects, the long-run distribution of choices between alternatives conforms in many cases to a rule that optimizes return only under special circumstances. At the same time, this rule has a deceptive appearance of rationality, so that it is easy to mistake it for a form of gradual optimization.

The second source of support is somewhat more speculative. Consider some typical examples of type B choices, or *distributed* choices, as we will hereafter call them:[1] expenditure rate on various non-durables; frequency of athletic exercise; rate of work in free-lance type occupations; allocation of leisure time; rate of savings (or dissavings); expenditures on lottery tickets, and other forms of gambling. When people express dissatisfaction about their choices, their discontent seems clustered around these sorts of distributed choices. For example, complaints that one is working too hard (or not hard enough), exercising too little (or too much), wasting time, overeating, overspending, and so on, are commonplace. Many of these anomalously suboptimal patterns of behavior have already been noticed by economic theorists, and have stimulated a burgeoning literature on models of "self-control" (Ainslie, 1975, 1982, 1986; Elster, 1984; Schelling, 1980; Thaler and Shefrin, 1981; Winston, 1980).

The next two sections of the paper spell out the basic theory we are proposing. The following section then applies the theory to "pathological" consumption patterns, and shows that one should find a general underinvestment in those activities that exhibit increasing average returns to rate of consumption, and an overinvestment in activities that have an addiction-like interaction between value and rate. The final section compares the theory with other approaches to suboptimal choice.

---

1. The boundary between type A and type B choices will vary across individuals. For example, for some people the savings rate is determined by a deliberate, perhaps automatically enforced policy, while for others it is the unintentional by-product of their expenditure rate.

Melioration: Basic Concepts

We begin with a standard problem in consumer theory. Consider a person lunching at the company cafeteria, five times a week. Unfortunately, there are only two items on the menu: pizza and a generic sandwich, which happen to have the same price. In a typical month with 20 workdays, for example, the person might choose pizza 13 times and sandwiches 7 times. If this distribution is stable from month to month, with the person dividing choices in roughly a 2 : 1 ratio, standard consumer theory implies that the *optimal* utility maximizing allocation is two to one in favor of pizza.

Now, this is not the type of choice that a normal person would fuss over at great length. But suppose that our consumer is determined to optimize, and to that end decides to keep a daily utility log in which the satisfaction derived from each lunch episode is written down. Figure 14.1 displays the entries for one month, with the height of each bar representing the introspected quality of a lunch, recorded on an interval scale (differences are comparable, but the zero point is arbitrary). Based on the information in the utility log, is there anything that the person can conclude about the efficiency of the 2 : 1 allocation? No. The log reveals that pizza is still a slightly better meal, in spite of its greater consump-

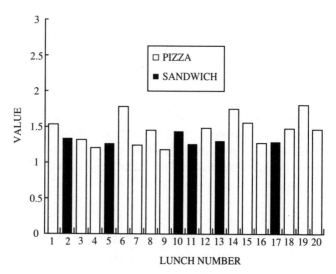

Figure 14.1   A utility log of satisfaction derived from lunch in a typical month.

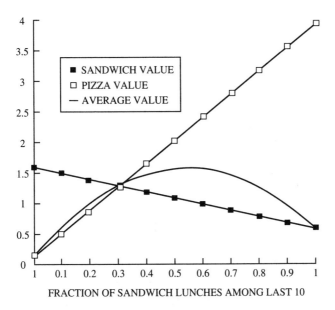

Figure 14.2 Value of pizza and sandwich lunches, as a function of last ten days' consumption.

tion frequency, and that consuming the same thing on two consecutive days usually (but not invariably) leads to a reduction in satisfaction. To test for optimality, the person would have to know the details of the hedonic mechanism that links present satisfaction with past consumption, and that mechanism is a black box.

But now take a look inside the black box responsible for this utility log. For this example, we postulated a hedonic mechanism that remembers the last ten choices, from which it computes the sandwich fraction, plotted on the x-axis of Figure 14.2. This is a "sufficient statistic" for consumption history. Since the fraction of pizza choices is one minus the fraction of sandwich choices, the pizza fraction can be read from the x-axis, too. The value of a sandwich on any given occasion diminishes with the number of times a sandwich was chosen in the last ten days, as indicated by the declining series of black squares in Figure 14.2. Likewise, the value of pizza, given by the white squares, diminishes with recent pizza consumption (reading the x-axis from right to left). Finally, to simulate the variability in a natural system, the values along the x-axis were subject to a random perturbation of up to 5 percent, so that, for example, three sandwiches out of the last ten would yield a fraction anywhere from 25 percent to 35 percent, with all values equally likely. This contributes to the fluctuations in Figure 14.1.

To determine a utility-maximizing pattern of choices in this situation, one would have to compare the average utility levels maintained by consuming pizza and sandwiches in different relative proportions. This average value, indicated by the solid curve in Figure 14.2, lies between the satisfaction levels obtained from pizza and from sandwiches—specifically, it is a weighted average of the two quantities, with the weights corresponding to the relative fractions of pizza and sandwich choices. The average value is maximized if sandwiches are selected about 50–60 percent of the time.

Although the example is exceedingly commonplace, it has several aspects that make it difficult for the individual to maximize utility. The optimal distribution could be derived in at least three ways: (1) guessing the incremental utilities of the next pizza and sandwich choice; (2) comparing the overall utility levels associated with various long-run allocations of choices; or (3) estimating the shapes of the value functions and computing the maximum (as in Figure 14.2). All three approaches are problematic. Let us consider them in turn.

Always choosing the item that has higher incremental value is an optimizing strategy, provided one knows the correct incremental values. Suppose, for example, that a person has been choosing sandwiches 40 percent of the time, and now wants to decide whether the next selection should be pizza. The difference in the incremental values of pizza and sandwich is composed of two parts, the first being the difference in satisfaction experienced at the time of consumption (about +.4 in favor of pizza, reading from Figure 14.2), and the second being a slight change in the values of the next ten meals. This second factor counts in favor of the sandwich selection (because the value of sandwiches is less sensitive to its own consumption rate), but who can tell offhand where the net balance of these two incremental value components lies?

Turning to the second approach, it is done in principle to attempt to compare the average returns associated with differing choice distributions. But to conduct this introspective exercise, one would have to mentally consume, say, a 40–60 mix of the two meals, and compare it to a 60–40 mix, and an 80–20 mix, and so on. Can a person discriminate among satisfaction levels produced by meal series that differ in the relative frequencies of items?

Finally, the third approach presupposes not only that a person knows the shape of the value functions—how value diminishes with consumption frequency—but also that the consumer recognizes the need to compute a weighted average of the value levels at different relative consumption frequencies.

The central conjecture we now make is that in problems of this general type people will not use any of these three methods to calculate the overall utility level associated with particular distributions of choices. Instead, they will choose as if guided by the following pair of rules:

(1) *Value Accounting:* Keep track of the average satisfaction (or value) received per unit invested on each alternative (where the unit of account might be a single choice occasion, a time interval, or a money unit).

(2) *Melioration:* Based on these value accounting calculations, shift behavior (choice, time, money) to alternatives that provide a higher per unit return.

If this process continues, we will observe a stable distribution of choices such that the accounted values in the long run of both alternatives are equal, or, as the other possibility, that only one alternative is chosen (see Thaler, 1980, for an early consideration of value accounting). In our example, the natural value-accounting scheme credits satisfaction to either a "pizza" or "sandwich" category. Figure 14.2 indicates that pizza will appear more tasty if it is consumed less than 70 percent of the time, less tasty if consumed more than 70 percent of the time. The intersection of the two value lines, at 30 percent sandwiches and 70 percent pizza, identifies the allocation at which a person who is practicing value-accounting and melioration will stabilize. However, this is not optimal, and the person would suffer a modest loss in overall value or utility relative to the maximum that the situation makes potentially available (at 60 percent consumption of sandwiches).

Implicit in the melioration model, then, are some assumptions about how a person summarizes information about satisfaction (utility) derived from alternative sources. Expressed in ordinary language, the model assumes that people are primarily able to assess how much they like specific activities or objects: whether they like a good book more than TV, sports more than classical music, Chinese or Italian food, and so on. What they have difficulty with, however, is in assessing the relative standing of entire distributions of activities or objects—whether a 20–80 mix of reading and TV is preferred to a 50–50 mix, for example. The source of the difficulty is that pairs of distributions are not concurrently available for comparison, unlike the objects or activities over which the distributions are defined. [A section describing the procedure and results of the experiment in Figure 11.2 of Chapter 11 (the Harvard Game) is omitted here.]

As for the general relevance of the finding, the procedure can readily be seen as a laboratory simplification of real-life economic situations. The decision-maker's past choices interact with the reward structure. Choices of restaurant, entertainment, product brands, minor investments or expenses (like clothing, decorative jewelry, repairs to an automobile or television set) are typically made without proper regard to the way they may, in turn, affect the reward structure by having been chosen. Effects like this can be created by economies of scale, by depletions of the reward resource, by time-dependent changes in the external environment and in motivational states, and the like, all of which, in turn, may be affected by past behavior.

General Definitions of
Value Accounting and Melioration

The results of this experiment and many others (cited below) prompted the following formal definition of the conditions that a choice distribution should satisfy for it to be a possible end result of a meliorating process. As stated earlier, our intended area of economic application is to the class of problems in which total expenditure on a commodity is an aggregate of many temporally distinct choices. This section uses the standard notation of utility maximization to explain how this model of choice differs from the standard model.

The basic variables, denoted by x_i, are the rates at which individual items, indexed by $i = 1, \ldots, n$, are chosen (that is, purchased) over an observation interval. The total choice rate is subject to a budget constraint, $\sum p_i x_i = k$, where k denotes income. A complete choice distribution is the vector of rates at which items are purchased, $x = (x_1, \ldots, x_n)$. A utility function $U(x)$ represents the individual's true welfare (in terms of utility) if his or her purchase rates are stabilized at x. While it can be said that the individual is aware of $U(x)$ in the sense that he or she experiences the current value of $U(x)$, it is not assumed that the individual knows which x is optimal, nor indeed which of two distributions has higher utility.

Instead, choice in this framework is driven by a complement of value accounting functions (or value functions, for short), $v_i(x)$, one for each alternative, that indicate how much value or satisfaction is perceived to be obtained (on average) from a single purchase of type i, when the choice rates are given by x. The only constraint on the value functions is that they must account for all utility obtained: $\sum x_i v_i(x) = U(x)$.

Given freedom to redistribute choice, a meliorating decision-maker will favor higher-valued activities at the expense of the lower-valued ones. The process of redistribution will then stop either at a corner distribution, where one alternative absorbs all choice, or at a distribution where all surviving alternatives have equal value per dollar (or per other behavioral investment). In the experimental literature, this distribution is referred to as "matching," inasmuch as all active alternatives yield equal returns per unit of behavior (or money) invested.

Like the original utility function, $U(x)$, the value functions define the relationship between tastes and consumption; in addition, however, they also reflect the manner in which a person interprets the information that the body and imagination provide. The analogy with an accounting system is quite apt here, since an accounting system also decomposes a single dimension of value (overall profit) into the per-unit profitability of individual products, which then must exhaust total profit.[2] However, two firms with different accounting systems might believe that their profits spring from different sources, which would lead the firms to behave differently. By analogy, it follows that two individuals might share the same "tastes," in the sense of having the same total utility function $U(x)$, but might interpret the source of their utility differently, by employing a different value accounting scheme. In that case, their choices and "revealed preferences" will, of course, not coincide. The extra measure of theoretical flexibility created by the value functions allows analysts

2. Faulhaber and Baumol (1988, p. 592) have observed that the influence of marginal analysis on business and government decision making has been minimal. Elsewhere, Baumol (1977) has listed several reasons why firms conduct their calculations in terms of average rather than marginal quantities, reasons that are relevant to our model of individual choice, as well. Perhaps the most important of these is that average quantities can be obtained on the basis of information about the firm's current operating levels, while marginal quantities, as Baumol (1977, pp. 34–35) notes, must intrinsically "represent answers to hypothetical questions."

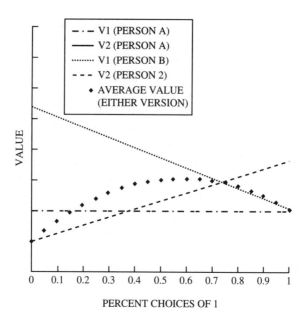

Figure 14.3 Two sets of value functions consistent with the same total utility function.

to relate apparent suboptimalities in consumer choice to particulars of the accounting scheme, as well as to other specific biases or illusions in value-attribution.[3]

Figure 14.3 illustrates this indeterminacy by displaying two different pairs of value functions consistent with the same overall utility function. From the earlier sandwich/pizza example, remember that the overall utility function can be thought of as the weighted average of the value functions for each of the two items separately. But two very different pairs of value functions can have the same average, as shown in Figure 14.3. The value function for commodity 1 is flat for person A, and declining for person B; the value function for commodity 2 increases more rapidly for A than for B. If they behave as meliorators, person A will choose commodity 1 only 18 percent of the time, while person B will choose it about

3. This chapter does not deal explicitly with the way the melioration principle can make contact with some of the paradoxes of choice arising in particular accounting schemes, or what is also called the "framing" problem (Tversky and Kahneman, 1981; Thaler, 1980), but see Herrnstein and Mazur (1987) and Rachlin, Logue, Gibbon, and Frankel (1986) for examples.

75 percent (at equal prices), even though the true preferences of the two individuals coincide perfectly, as shown by the fact that the same curve of diamonds represents a weighted average for both A's and B's preferences. Again, the analogy with corporate accounting systems is quite tight. Two firms with the same production function may have different policies for allocating overhead or advertising costs, and as a result would reach different production decisions.

If value functions are allowed to vary freely, it becomes possible to explain any behavior whatsoever. Thus, if value functions are to provide genuine insight into behavior, they should codify an accounting scheme that is psychologically plausible. The early experimental demonstrations of meliorating behavior provided a good example of what we mean by psychologically plausible accounting schemes. These experiments trained animal subjects to make long series of choices between two gambling devices, each of which would occasionally deliver some reward—usually a small amount of animal feed. The probability of reward on each alternative was not constant, from one choice to the next, but increased with the time since the last response on that alternative, thus implementing a nonlinear reward-response technology.[4]

Several hundred experiments have confirmed the prevalence of meliorating or matching behavior in human and animal behavior with respect to a "natural" value accounting rule, according to which the current value of an alternative equals the ratio of benefits to costs, i.e., the rewards per responses for that alternative. A review by Williams (1988, p. 178) in the *Stevens' Handbook of Experimental Psychology* states that: "The generality of the matching relation has been confirmed by a large number of different experiments. Such studies have shown matching, at least to a first approximation, with different species (pigeons, humans, monkeys, rats), different responses (keypecking, lever pressing, eye movements, verbal responses), and different reinforcers (food, brain

---

4. In the notation that has become standard for experimental psychology, the central result of these experiments was stated as a matching principle (Herrnstein, 1970):

$$\frac{B_1}{B_1 + B_2} = \frac{R_1}{R_1 + R_2},$$

where B_i and R_i respectively refer to the rates of responding and of reward collected on a pair of alternatives, $i = 1, 2$. The rates are simply the number of events (responses or rewards) over the duration of the experimental session. The equation implies that $R_1/B_1 = R_2/B_2$, or that the same fraction of responses is rewarded on both schedules.

stimulation, money, cocaine, verbal approval). Apparently, the matching relation is a general law of choice."

An efficient allocation of effort, one that extracts the most reward for a given rate of work, would require that the marginal benefit per marginal increase in choice rate be the same across all sampled alternatives.[5] At the matching point, however, the average benefit per dollar is equalized. The resulting inefficiency can be likened to an *externality*, in that the meliorating individual ignores the effect of a change in choice rate on the values of the sampled alternatives (including, importantly, the value of the alternative in question).

Among possible meliorating outcomes some will be unstable, in the sense that a small—perhaps unintentional—increase in the choice rate for a particular alternative will cause that alternative to increase in value relative to the others, thus initiating a further increase in its choice rate. In contrast, there are other cases where any brief period of experimentation with the new distribution will cause the alternatives to appear worse, so the original meliorating outcome is stable in that sense. In the delay experiment described in the earlier section, for example, excursions away from choice 1 will appear to make the delays longer, rather than shorter.[6]

Although stable meliorating choices are not necessarily optimal, they do exhibit a semblance of rationality. If the values of all alternatives are constants, independent of choice rate, then the unique stable matching point is at exclusive preference for the alternative that offers high-

---

5. What does a "first-order" optimality condition look like for a meliorating decision maker? Noting that the true utility function is a sum, $\sum x_i v_i(x)$, we can write out the marginal utilities as:

$$\frac{\partial U}{\partial x_i} = v_i + \sum_j x_j v_{ji},$$

where $v_{ji}(x)$ is the partial derivative of $v_j(x)$ with respect to x_i.

6. A necessary condition for stability is that the value per dollar on each alternative is no greater than the value per dollar obtained from all alternatives:

$$\frac{v_i}{p_i} \leq \frac{\sum_j x_j v_j}{\sum_j p_j x_j}, \quad \text{for all } i,$$

or otherwise alternative i would be chosen more often. The issue of stability can become fairly complex when there are three or more alternatives, and is discussed in detail elsewhere (Herrnstein and Prelec, 1992b).

est value per dollar (more precisely: all corner distributions satisfy the matching law, but only the best one is stable). Even if values are dependent on consumption, the person will "correctly" adjust consumption in response to a uniform improvement or deterioration in the quality of a single alternative in the way that standard utility theory would predict. A more entertaining television program will fetch a larger audience; likewise, the information that cigarettes are harmful will cause people to reduce their smoking rate, on average, and so on. As a result, much of the behavior of a meliorating person will appear rational on the surface. However, such examples do not discriminate between optimization and weaker forms of reward-seeking behavior, such as melioration (an analogous point was made by Becker, 1962, in a different context).

Pathologies of Distributed Choice

Although the meliorating process has a flavor of rationality to it, the meliorating consumer will generally not home in on the optimal distribution, and may, in some situations, settle at the worst possible distribution. Do similar examples of poor, yet stable, choice distributions exist outside the laboratory?

It is a commonplace that certain patterns of consumption may diminish personal welfare. Most people agree about extreme examples, like severe drug addiction, perhaps because of the clear testimony of former victims. But such obviously disastrous consumption patterns are surrounded by a penumbra of less clear cases. For example, television has been labelled addictive by some, as have gambling, shopping, athletic exercise, personal relationships, and even plain work (Becker and Murphy, 1988). If the concept of consumption is broadened to cover choice of lifestyle, or character, we would find a host of tenacious yet unsatisfying patterns of behavior, ranging from the traditional seven deadly sins (lust, wrath, avarice, pride, envy, sloth, gluttony), to those that modern psychiatry recognizes as personality defects (Ainslie, 1982).

These phenomena share a common element: they are all instances of distributed choice, as the term is defined here. A person does not normally make a once-and-for-all decision to become an exercise junkie, a miser, a glutton, a profligate, or a gambler; rather, he slips into the pattern through a myriad of innocent, or almost innocent choices, each of which carries little weight. Indeed, he may be the last one to recognize "how far he has slipped," and may take corrective action only when prompted by others.

In its most egregious form, the slippery slope of distributed choice leads to addiction, which is to say, a devastating level of overindulgence in some commodity or activity. To the extent that economic theory has addressed the problem of addictions, it has done so in essentially two ways. One approach, initiated by Stigler and Becker (1977) and further developed by Becker and Murphy (1988), builds a taste-changing mechanism into a global, multi-period consumption function, so that the marginal substitutabilities among various commodities change as a function of consumption history. The demand functions that result from a once-and-for-all maximization of the intertemporal utility function will exhibit changes over time (in demand elasticities, for example) that mimic "harmful" and "beneficial" forms of addiction. Because tastes are fathered by a global utility function, each consumer, no matter what his or her consumption pattern, is still at a personal optimum, according to this theory, and would presumably benefit from a reduction in price, or increased general availability, of any commodity or substance, no matter how lethal by ordinary standards.

The second (but earlier) approach, associated with the work of Pollack (1970) and von Weizsäcker (1971), among others, assumes that the taste-change process (also called "habit-formation") is essentially opaque to the consumer. In von Weizsäcker's two-period model, for example, a consumer myopically maximizes current-period preferences, which are themselves conditioned by consumption in the previous period. However, this work draws a sharp separation between a type of taste-change to which the person adjusts optimally, namely the taste-change that is encompassed within the one-period preference structure, and a second type of taste-change, across periods, which the consumer ignores completely. It is not made clear in this theory why the process of taste change encountered in addiction should be treated differently from the ordinary taste-change that occurs when, for example, one eats fish too many days in a row.

The meliorating theory of distributed choice developed here models choice as potentially suboptimal whenever tastes ("values," in our terminology) are affected by rate of consumption. What distinguishes more or less non-problematic choices, such as allocating the food budget, from clearly problematic ones, like addictions, is not that the typical individual has a technique for maximizing the former, which he fails to apply to the latter, but, rather, that the value functions for the first category produce a stable matching point that is relatively efficient (in terms of maximizing utility), while those for the second category do not.

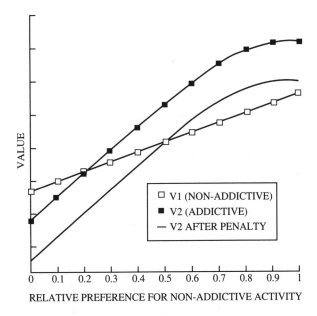

Figure 14.4   An addictive configuration of value functions.

Figure 14.4 displays a configuration of value functions for the choice
between a non-addictive and an addictive behavior. In this case, the resulting equilibrium can be markedly inefficient. Notice that the value
function for choosing higher proportions of the non-addictive activity
(open squares) is steadily increasing, while the value function for choosing the addictive activity (filled squares) starts at a high level and then
decreases (reading right to left). The result of meliorating behavior will
be an equilibrium where a great deal of time is devoted to the addictive
activity, even though a higher utility level could be reached by sticking
exclusively to the non-addictive behavior. Consider two possible interpretations of the pattern in Figure 14.4.

One possibility is that alternative 1 represents an activity whose value
intrinsically depends on its own rate of choice, and the value is increasing with rate. This type of value function would be characteristic of
skilled activities (like music or sports) which provide greater satisfaction
if maintained at higher rates. It is a necessary property of meliorating
equilibria that any activity whose value increases with consumption will
be underconsumed. To understand this point, consider a move away
from the equilibrium in the direction of more 1-choices (this would be a

rightward movement in Figure 14.4). If the equilibrium was stable, then this change in v_1 will have to be accompanied by an even greater positive change in v_2, so that behavior is pushed back to equilibrium. But both alternatives have increased in value! As a result, total utility must increase with increased allocation to 1; hence the original equilibrium was inefficient.

A second possibility is that alternative 1 does indeed have the usual property of diminishing marginal value with consumption, but this effect is swamped by a negative impact of alternative 2 on the value of 1. Consuming the second alternative, in other words, destroys the satisfaction produced by engaging in alternatives complementary to it. This seems to correspond to the common understanding of harmful addiction, as a process in which consumption of the addictive substance casts a shadow over other pleasures. (A more complete account of addiction in the melioration framework is presented in Chapter 9.) Students of opiate addiction often observe, for example, that addicts lose their appetite for food, making simple malnutrition one of the primary health risks for addicts. In other words, as the consumption level for v_2 (heroin) rises, the value of v_1 (food) declines. Or as another example, the habitual heavy use of alcohol erodes the gratifications to be had from family, work, and even ordinary physical pleasures as one's health deteriorates.

The sharply diminishing value of v_2 as a result of its own consumption is likewise consistent with the phenomenology of addictive substances. For the usual addictive substance, high rates of consumption typically lead to the development of "tolerance," which is to say, a shrinking harvest of utility from a given quantity of the commodity. Richard Solomon's opponent-process theory of acquired motivation (Solomon and Corbit, 1974) describes how positive pleasures at a low rate of consumption are transformed into the avoidance of the agony of withdrawal at high rates. The remaining pleasures are then minimal.

At the eventual stable meliorating equilibrium, the value of the addictive alternative has itself almost evaporated, that is, it has been reduced to the value level of a low amount of the complementary activity. In the case of overeating, for example, this state describes a person who eats at a more or less continuous rate, not allowing true hunger to develop. As long as food is readily available, such a person would not experience eating as especially pleasurable, a finding which has some support in the research literature on obesity (Schachter, 1971).

In this analysis, a nonfinancial penalty, hindrance, or a substitution of an inferior commodity for the original alternative 2 (like methadone for heroin, for example) may improve individual welfare.[7] Effectively, by offering an addictive alternative that is less attractive, it would amount to moving the v_2-curve down, which would move the meliorating equilibrium to a higher value level. This is illustrated in Figure 14.4 by the intersection between the non-addictive alternative and the addictive alternative labelled "v_2 after penalty."

The element common to both interpretations of addiction is an underconsumption of the first alternative; the difference is that in the first case the suboptimality is accounted for by a peculiarity of that same alternative, namely, its increasing returns to consumption, while in the second case the culprit is addictive alternative 2, because it destroys the value of the complementary activity. Using the terminology of Stigler and Becker, we would say that meliorating persons generally underinvest in beneficial addictions, for which training and exposure build up "enjoyment capital," and overinvest in harmful addictions, for which enjoyment capital depletes rapidly. Stigler and Becker's choice of "beneficial" and "harmful" as labels is quite revealing, given that their theory does not allow for any suboptimality in consumption patterns. Melioration theory explains why the errors people make in executing an optimal plan cumulate in different directions for so-called beneficial and harmful addictions. Maximization theory, which allows no systematic errors, does not.

The value functions that lead to meliorating behavior may have the shape they have in Figure 14.4 for many reasons. Exogenous or endogenous chemical agents may alter reward centers in the brain in a particular way, one may learn how to enjoy ("acquire a taste for") a certain class of rewards, different rewards may have different time horizons associated with them, and so on. But if, for whatever reason, the resulting structure conforms to that outlined here, the result is a pathology of choice.

Conclusion

Most social scientists have certain favored explanations about the most prevalent causes of inefficient or dramatically suboptimal behavior, intuitions, that sometimes give rise to full-blown theories. In this concluding section, we would like to clarify the relation of our approach to some other approaches to suboptimal choice behavior.

---

7. The effect of a *financial* penalty, such as a tax, is ambiguous, because of the income effect.

Satisficing

According to the satisficing concept (Simon, 1955, 1956), economic agents fail to maximize utility perfectly, but they do maximize "well enough," as defined by their own aspirations. In our theory, agents may sometimes maximize, but in other circumstances, their choices may be suboptimal, far worse than any conceivable level they may originally aspire to. One could say that the critical difference between melioration and satisficing is that melioration describes an endogenously changing aspiration level, which is the mean value obtained from all sources at a choice point, while satisficing assumes an aspiration level that is externally set, typically by a larger institutional context.

Excessive Discounting or "Myopia"

The behavior of a person with a high discount rate would in some respects resemble that of a meliorator, in that such a person would on each occasion choose the alternative that offered the highest value, but nevertheless be choosing suboptimally in a broader time frame. Time discounting is often invoked as an explanation of this same pattern of over- and under-investment: A person with a sufficiently high discount rate would simply not care about acquiring the skills necessary to enjoy the so-called beneficial addictions, nor would he worry about the future penalties of harmful addiction. Unusually sharp time preference may indeed be a contributing factor to addiction, but it is conceptually distinct from melioration.

First, consider the psychological cost of a single drug-consumption episode as composed of two separate factors. The cost—an increased tolerance level, a degradation of other activities—is delayed in time compared to the "benefit" of drug use, which may be instantaneous; this is the time-discounting perspective. In addition, the impact of that single episode on the tolerance level and on other activities is small, and probably difficult to imagine even if one were perfectly informed: that is the perspective peculiar to distributed choices. The total cost of the episode, one should remember, is the aggregate, appropriately discounted, future reduction in pleasure caused by the incremental shift in tolerance, as well as a degradation in the future value of competing activities. Even in the absence of time discounting, we claim, people would still be vulnerable to addiction if they were unable to assess correctly and compute such costs.

It seems to us that addiction is more often a process of invidious slippage than a cool decision adventitiously supported by a high discount rate. Educational materials designed to discourage addiction typically do not work on altering time-preference, but on persuading people to think of addiction as an all-or-nothing decision. They warn that you cannot maintain a reasonable or moderate rate of drug consumption, that it is a case of all or nothing at all. The task of the educator is, in short, to get the person to act as if the choice is a one-time event, rather than distributed.[8]

Another way of noting the difference between discounting and melioration is to remember that a myopic person should shift toward the optimal allocation as the time scale over which the choices are made is systematically reduced. Melioration, in contrast, makes explicit reference to time only insofar as it distinguishes between distributed choice and nondistributed choice. Given that choice is distributed, it makes the same prediction irrespective of whether the choices are made once a week, or once every few minutes. Certainly, the experimental results described earlier[9] cannot plausibly be attributed to time discounting, since the consequences of single choices were typically deferred for less than a minute, and the entire experiment for a subject did not last much more than 20 minutes.

Ignorance or Incomplete Learning

A person who believes mistakenly that values remain constant, independent of choice rate, would behave exactly like a meliorator. However, this does not imply that melioration is necessarily produced by the assumption of constancy. It is possible that a person is aware of the relationship between choice rates and values, but that he sees no reason to doubt that a melioration-like process is nevertheless an efficient behavioral choice rule. When people are asked to imagine choosing among alternatives whose values interact in a particular way with choice rate, and they have been told what this interaction is, their answers typically display the same kinds of suboptimalities as are observed in experimental simulations of these situations (examples in Chapter 11 and Herrnstein

---

8. A discussion of how to avoid or escape addiction within the melioration framework is offered in Chapter 9 and Prelec and Herrnstein (1992b).
9. See Figure 10.2. [Editors' Note.]

and Mazur, 1987). People can be given the information they need to maximize and fail to use it correctly.

Taste Change

In the standard utility-maximization theory, changes in the marginal rates of substitution, embodied in indifference curves, constitute an implicit theory of taste formation in response to consumption levels. Indeed, any theory that includes a concept of subjective utility or value must infer the subjective entity, which is in principle not directly observable, from some observable aspect of behavior. The difference between the present theory and the standard one is, then, not in the fact that it postulates taste changes, but in the hypothesis it uses to connect subjective utility or value to behavior (Prelec, 1982, 1983). The present theory employs the hypothesis of melioration, rather than maximization. It is this difference, rather than the particular suppositions about taste changes, that sets the theories apart.

The notion of melioration echoes ideas at other levels of analysis (see Chapter 10). At a more inclusive level, consider an n-person, noncooperative game (Schelling, 1978). Each player will adopt the strategy most favorable to him, which, because of various externalities, may not be the optimal distribution of strategies for the players as a group. At a much smaller level of analysis, evolutionary biologists have been alerted to the importance of "frequency-dependent selection," the recognition that the selective fitness of a gene may depend on its proportion in a population. Genes in competition are, in other words, subject to the effects of externalities. As a result, the equilibrium points for genetic competition may not coincide with the optimum for the species as a whole (Dawkins, 1976).

What is new here is the suggestion that the familiar notion of externalities producing nonoptimal equilibria applies to the analysis of individual human (or animal) behavior. One could say that the difference between maximization and melioration is that melioration contains a weaker concept of personal identity. Within the maximization framework, the alternatives are, in principle, so perfectly articulated with each other that they lose their individuality. Only the utility of the whole choice bundle matters, rather than the utility attached to its constituents. Within the melioration framework, the alternative behaviors are in competition with each other, vying for the organism's investment of time, effort, and so on. Beyond the dynamics of this competition itself, no one is in charge.

15 | Preferences or Principles: Alternative Guidelines for Choice

DRAŽEN PRELEC AND RICHARD J. HERRNSTEIN

"Profound moral conviction was the basis of Mr. Gladstone's political influence," remarked Bertrand Russell, explaining: "Invariably he earnestly consulted his conscience, and invariably his conscience earnestly gave him the convenient answer." From the perspective of a rational agent, the humor in Russell's joke ought to be cryptic, but, of course, it is not, especially to those whose conscience is not as obliging and versatile. There is something odd about the suggestion that one's conscience reliably serves one's self-interest, but why? At the level of common sense, we all recognize that actions are shaped by such things as values, rules, vows, obligations, and principles, which are in part our own creation but which can also appear to us as an external constraint, over which we have little control. Can that idea be incorporated into an empirically sound, but theoretically consistent, theory of choice?

In this essay, we would like to suggest the importance, for the social sciences generally, of having a theory of human nature and action within which the paradox inherent in Gladstone's decision-making style would find some natural explanation. We do not have such a theory in any rigorous sense of the term, and we do not think a fully satisfactory one is available, despite the efforts of philosophers and others who have

Originally published in R. J. Zeckhauser (ed.), *Strategy and Choice* (Cambridge, Mass.: MIT Press, 1991). An earlier version, by Prelec alone, was presented at the session "Socioeconomics: The Roles of Power and Values" at the annual meeting of the American Association for the Advancement of Science, San Francisco, January 14–19, 1989. The version here has had the benefit of many useful comments and suggestions from Thomas Schelling, Richard Zeckhauser, and other members of the seminar on public policy at the Kennedy School of Government at Harvard University. Jonathan Riley provided helpfully skeptical comments. We are much in their debt. Many of the ideas presented here were developed as resident fellows, in 1988–89, at the Russell Sage Foundation, to which we owe many thanks for supporting our work.

thought long and hard about these matters. Hence the essay will have the more modest aim of showing the need for a *psychology of legal/moral reasoning and interpretation* that would describe how people use principles in decision making, as well as how principles are integrated with rational cost-benefit calculations.

Rational choice theory seems to be readily reconciled with the existence of explicit *moral* principles, at least for those cases in which individual compliance can be monitored by the community. It is well understood that a collection of individually maximizing agents may often fail to find a collective optimum when confronting prisoners' dilemmas or common grazing lands, if they are not guided by explicit rules. But what of *prudential* rules of self-management (Schelling, 1985)? If we are rational individually, why do we need anything more to guide us? If we prefer to weigh less, be in better shape, or stop drinking, why do we not just do so? The fact that prudential rules exist indicates that the tendency to form, and follow, rules requires some explanation at the level of individual psychology.

Introspection suggests that moral and prudential principles tap a common psychological mechanism, although the object of concern is different. We may feel guilty or remorseful for having overslept or overspent or broken a diet, just as for violating a law or a duty or a social obligation. We may feel virtuously good for declining the chocolate cake or for contributing to charity. Violating a convention for which there is no corresponding internalized principle probably engenders only embarrassment, not guilt, and that only when our violation has been detected by others. Conforming to a convention lacking an internalized principle may confer no further benefit than the avoidance of embarrassment, and no warm glow of virtue.

The next section of the essay defines rules or principles and explores their possible function. We then discuss why rule-governed behavior resists assimilation to the rational cost-benefit model. This same issue is taken up again in the following section, but now with respect to so-called divided-self models, which view the individual agent as a collection of noncooperating but individually rational subunits. The essay concludes with a short summary of the basic differences between rational and rule-governed action.

Systematic Biases in Cost-Benefit Calculation

By one definition, a principle does not *supplement* ordinary cost-benefit analysis, but rather *replaces* it (Etzioni, 1988). In other words, by invoking a principle one does not add another consideration onto what Janis and Mann (1977) call the "decisional balance-sheet"; instead one discards the balance-sheet altogether, at least provisionally. There is nothing wrong with thinking about the costs and benefits of a marriage before the vow[1]; afterward, such calculations probably foreshadow the end of the marriage.

In our view, a behavioral policy in regard to some action or class of actions is a rule or principle[2] if, and only if, it overrides cost-benefit calculation with respect to that action. Within the normative economic theory, the approach is to fit rules and principles into the utilitarian calculus. Our claim, in contrast, is not that the standard approach is unworkable, but that the issue is clarified by proceeding as if there are two distinct ways of deciding on a course of action, only one of which involves the utilitarian calculus. The second method is used not merely because the calculations are often hard to do, as they certainly are. Even when the calculation is easy, as subsequent examples will show, we often invoke a guide to behavior that abjures doing the mental arithmetic of cost-benefit analysis.

Perhaps a clarification of the word *rule* might be useful. Within the tradition of "bounded rationality" theorizing in economics, decision-making rules have been viewed as a necessary response to the daunting complexity of the decision problems faced by both individuals and economic organizations. The tradition probably can be traced to Herbert Simon (1957), and shows its influence today in the work of Heiner (1983) and Nelson and Winter (1982), among many others. Simplifying somewhat, we could say that a rule, in this view, is a behavioral policy that produces satisfactory results in the long run, without burdening the decision maker with case-by-case cost-benefit calculations. This is what Etzioni refers to as "rules of thumb" (1988, chapter 10), and Nelson and Winter (1982) as "routines."

Our focus of attention is on the sort of personal rules that arise, not to routinize calculation, but to overcome weaknesses of our natural cost-benefit accounting system, weaknesses of which we are aware and for which the rule or principle serves as a partial antidote. For example, parents who are not themselves religious may try to inculcate

---

1. In the vow, we may foreswear calculating "sickness," "health," "richer," "poorer."
2. Richard Zeckhauser has pointed out (personal communication) that the term *principle* seems to refer to deeper guidelines for conduct, which require additional elaboration before they can be applied in a given instance. Rules, on the other hand, are relatively unambiguous directions for action. That distinction in usage will not be observed here, although it may prove to be useful in more detailed analyses.

religious belief in their children, because they feel that the principles of a religious life lead believers to make better choices than nonbelievers do. Some parents may inculcate belief because they think that *society* is better served by believers, but our point is that some parents may do so because they think their *children* are better served by being believers.[3]

Such rules need not be imposed by parents or other agents who have our interests in mind; they may also be "constructed" by ourselves as we see the need for them. We may observe, for example, that our alcohol drinking is creeping upward—our taste for alcohol is pushing us beyond what we perceive as an optimal level of consumption—and decide to adopt a one-drink-a-day policy, which may be an improvement without being quite optimal. We may realize the risks of drinking, but find ourselves at times unable to act sensibly in light of our realization. In the final section we discuss briefly this contrast between internally and externally imposed rules or principles, but until then, the distinction will be set aside.

Rules or principles cannot be a perfect antidote under all conditions, for if they were, then they would be equivalent to a perfect cost-benefit analysis, which is, by assumption, not possible. One might expect to find rules proliferating in exactly those choice domains where a natural utilitarianism produces results that we do not find satisfactory. Rules or principles presumably improve performance over the results of our faulty cost-benefit analyses, but not to virtual optimization.

At least three different forms of trouble afflict ordinary cost-benefit analysis. Each can be described as a biasing asymmetry or mismatch in the disposition of the costs relative to the benefits that flow from an action. The tacit assumption is that a decision maker must choose between taking or not taking a course of action. The consequences of not taking the action are, for simplicity's sake, assumed to be a fixed standard of reference against which we compare the results of combining the costs and benefits of the action. We omit here, as we do elsewhere in this essay, the complexities of strategic interactions, in which present choices affect the outcomes of future choices through their influence on other agents. What difficulties may we encounter in this calculation?

---

3. We refer here to the benefits from choices made as a believer, which would have been different for a nonbeliever, not the benefits (e.g., respect, deference, etc.) that believers receive from other people by virtue of being believers.

(1) *Temporal mismatch,* in which the cost and benefit are separated by a substantial time interval.

(2) *Saliency mismatch,* in which one element of the cost-benefit pair is vivid and easy to imagine, whereas the other is not.

(3) *Scale mismatch,* in which one element of the cost-benefit pair is perceived to have impact only in an aggregate sense (that is, only if the same action is repeated many times, or on a larger scale).

We will refer to these mismatches as asymmetries of time, saliency, and scale, respectively. In all three cases, cost-benefit analysis fails because it assigns the wrong weights to one or the other, or both, of the elements of the pair, given the person's long-run valuations. The excessive discounting of future consequences is amply demonstrated by both casual observation and experimental research (Ainslie, 1975), hence the asymmetries of time. As regards saliency, the importance of vividness has been demonstrated in the domain of probability judgments by Tversky and Kahneman (1981), and it seems likely that a comparable bias would afflict judgments about utility and value.

The third mismatch, asymmetries of scale, which could also be described as the "drop in the bucket" phenomenon, has been analyzed by Herrnstein and Prelec (see Chapter 14) in the context of so-called *problems of distributed choice.* There are situations in which the economically significant variables are aggregates of many temporally distinct decisions, each of which, individually, has little impact (rates of cigarette consumption, frequency of exercise, or social interaction, etc.). But, to the extent that decision makers ignore these impacts, their choices may in the long run be predictably suboptimal.[4]

The first, and most critical psychological function of a rule, according to the line of thought being pursued here, is to *disengage* the cost-benefit calculus, because it is subject to distortion by one or another of the asymmetries. If this is true, then we should be able to trace specific rules to specific (possibly multiple) mismatches in the cost-benefit accounting. Here are some examples.

Rules Pertaining to Health and Personal Safety

Decisions that affect health are clear examples of distributed choice (or scale mismatch), because the cost of engaging in some unhealthy or risky

---

4. For a further discussion of the theory and relevant experimental evidence, see Chapter 14, Herrnstein and Vaughan, 1980, or Prelec 1982.

action is negligible unless the action becomes a permanent pattern or lifestyle. Try, for example, to evaluate the cost or the benefit of

(1) one cigarette;

(2) one slice of chocolate cake;

(3) one car trip without a seat belt;

(4) one visit to the health club;

(5) one more day at the beach, away from work.

In some of these examples, a temporal mismatch is also present (e.g., the health risks from smoking), but it is conceptually a distinct problem, and one that may not be of primary importance. Consider the seat-belt case. The cost, if it appears, is instantaneous; it is also vivid (the saliency mismatch does not apply). Furthermore, if you survive the trip, there are no insidious aftereffects—the balance sheet is cleared. Hence the problem is primarily one of scale.

Presumably, the cost-benefit problem in the seat-belt case could be tackled either "in the small" (i.e., for the one-shot decision) or "in the large" (i.e., whether one should buckle up always, or never). Neither format is convenient for a cost-benefit analysis, because of the mismatch in scale. In the large problem, the probability of harm that follows from a no-buckle-up policy can perhaps be appreciated, but how does one aggregate, across hundreds of trips in the car, the modest comfort increases garnered by not buckling up? In the small problem, the comfort can perhaps be appreciated, but the tiny probability of harm in one trip has little subjective meaning. The aggregate benefits of seat-belt use outweigh those of not using them, so, if we could correctly solve either the long or the short cost-benefit problem, we would use the belts.[5] Adopting a policy signifies that we are disinclined to solve the problem in terms of costs and benefits.

It could be argued that even though the stakes "in the small" are difficult to grasp, nothing prevents one, in principle, from solving the large cost-benefit problem and, as a result, deciding always to use seat belts. But although subjectively we may feel that we have decided "once

5. More likely, we would use the belts on some occasions and not on others, depending on how comfort and risk measure up. The very fact that we seize on using them always (or, for some people, never) suggests that a rule, rather than a cost-benefit calculation, is guiding behavior.

and for all" to use seat belts, objectively we are not given the opportunity to make such a choice. The objective situation presents us only with choices "in the small," each covering precisely one trip; if it were possible to buy a car that would not start with seat belts unbuckled, then one could choose a policy. As it is, a person may *decide* on a policy, but that is no guarantee that he or she will indeed follow it, when confronted with the reality of one trip at a time.

For people who regularly use seat belts, it may seem odd to categorize this activity with such notoriously conflictual matters as too much chocolate cake or smoking cigarettes or chronic absenteeism from work. Regular seat-belt users may want to distinguish between policies adopted with little or no internal fuss, and those we continually struggle to maintain against the temptations of defection. But, on the other hand, not everyone uses seat belts regularly and effortlessly. To them, the seat-belt issue may naturally sort itself with chocolate cake or absenteeism. Such questions as the relative universality of a particular decision-making conflict or the intensity of the conflict are important and deserve consideration. But, here, we will have little more to say about them (the final section comments on the motivational force attached to principles), beyond observing that conflicts may be more or less common, and more or less intense, and still exemplify scale mismatches.

Rules That Build Up and Maintain Character, or "Personal Identity"

People derive satisfaction from their internal moral standing or, more generally, from the subjective correspondence between their self-image and some ideal. But maintaining such an image is usually a problem of distributed choice. A character trait, for example, is the product of many separate decisions, any given one of which could have been reversed without major impact. It is probably true that a person can consider himself honest, generous, brave, in spite of a few deceptions, or occasional acts of cheapness or cowardice. (However, is this a truth for a moralist to advertise?) For example, the choice between truth and falsehood, in the small, pits the often clear and instantaneous benefits of a lie against the minute blemish to one's self-image of truthfulness incurred from a single lie. The problem of distributed choice is that the outcome of this decision may tip toward steady lying even if one believes that "honesty is the best policy."

Rules for Spending or Saving Decisions

Ordinary overspending may often be primarily due to a temporal mismatch—the expensive magazine picked up at the check-out counter, which you bought impulsively and would not have bought had you been making the decision a few moments before you reached the end of the check-out line. But some forms of overspending are probably more a matter of saliency mismatches. In "home shopping clubs," for example, television viewers are shown products that can be ordered by phone. It is probably clear to anyone that money spent to purchase a product (for, say, $19.95) will not be available to spend on something else—there will be a loss, in other words, of $19.95 in general purchasing power. But where, specifically, will the sacrifice be felt? Most people do not know. This creates an initial bias for overspending—the benefit is concrete and visible, while the cost is an unspecified and abstract corollary. The bit of consumption that will be canceled by the purchase is not present to the mind, to make its case felt.[6]

The success of home shopping clubs cannot be attributed to the immediacy of reward (i.e., the time mismatch does not apply). Products purchased over the telephone may not arrive for weeks, possibly well after the charge has been debited against one's account, given the vagaries of billing and shipping. Nor could one argue that the decision to buy is driven by the immediate *expectation* of using the product, because the loss in expected purchasing power is also realized immediately.

Undersaving, in general, may reflect the simultaneous operation of all three mismatches. First, the benefits of savings are delayed; second, the benefits of savings may be imagined at the level of aggregate future income, which is not appreciably affected by single acts of current consumption; third, the benefits are less concrete, given the uncertainty surrounding one's future circumstances.

Prohibition against Crime and Self-Abuse

Pure examples of a temporal mismatch are hard to find; they would be cases in which the delayed consequence was clear and hinged on a single action. Some criminal acts probably fall in this category (Wilson and

---

6. There is a parallel here with the so-called invisible lives problem in public project risk-management. It is notorious that life-saving funds are more easily appropriated if they will benefit a visible group than if they benefit an anonymous sample from the population. For example, expenditures for traffic safety are typically lower than the level that would be justified by a lives-saved analysis.

Herrnstein, 1985), especially "crimes of passion" that are committed in full knowledge that punishment is probable. Other examples might include drug, alcohol, or sexual binges, in which the negative hangover phase is fully expected, but discounted.

When Opportunities and Information Have Negative Value

Our hypothesis is that rules or principles take over in situations where cost-benefit analysis systematically fails at the personal level. But can the use of rules itself be rationalized by a more complete inventory of individual preferences, one that would capture a taste for following principles? We think not. Without going into technical detail, and omitting the complexities that arise in strategic interactions between rational actors, the modern conception of the rational agent derives from three fundamental psychological assumptions:

Assumption (1): A person's preferences over a given set of outcomes of choices are independent of how the choices are constrained.

Assumption (2): All aspects of individual psychology bearing on choice can be resolved into matters of taste (preferences) or knowledge (information).

Assumption (3): A person chooses the most preferred outcome, given the information at his or her disposal.

From these three postulates flow two elementary theorems about how changes in opportunities, information, and preferences should affect the utility or welfare derived from a choice situation:

Theorem (1): Expanding opportunities can only increase personal welfare in the utility-maximizing sense.

Theorem (2): Having more information can only increase personal welfare, similarly defined.

It is important to recognize the utter generality of these deductions, at the purely individual level (i.e., in the absence of multiperson externalities). They are valid without regard to a person's tastes, or the extent to which he understands the choice situation; they constitute the empirical rock bottom of the rational model. Thus, it is significant that decisions involving principles will often produce violations of the two theorems.

Costly Opportunities

Suppose that you are given an opportunity to do something, which, upon consideration, you decline. If the rational model is correct, your welfare should be unchanged—you are exactly where you were before the declined offer. And indeed, insofar as *tastes* are concerned, the prediction holds: if you are about to order steak from a restaurant menu when the waiter informs you of a special unlisted dish, you can reject this new possibility without impairing your enjoyment of the steak.

But now suppose, instead, that you are an unemployed football player, invited to join one of the "replacement" teams during the 1987 football strike. If you reject the offer, the regret over lost income may make you suffer. But if you accept the offer, you may regret having violated a personal rule against strike-breaking and may wish the situation had never arisen.[7] The clearest challenge to rational choice theory is posed by the case in which the individual declines the offer, but feels worse off for having had the opportunity. Such phenomena are recorded in speech, when we say things like, "It is a tempting opportunity, and I'd like to take it, but I regret I cannot accept." While such expressions of regret are sometimes merely conventional, in many cases they are sincere. Yet, if the opportunity failed to pass a comprehensive cost-benefit analysis, why regret declining it?

If the job offer is perceived as a temptation, and not just something one either wants to accept or reject, then it can be neither refused nor accepted without cost. An option to do or not to do something becomes a temptation when it presents an opportunity to break a personal principle. For this reason, it is considered unkind, if not worse, to offer cigarettes to a recently converted nonsmoker, or cakes to someone on a diet, or drinks to an ex-alcoholic, or a sexual proposition to someone else's faithful spouse. Some social and legal institutions specifically shield people from temptations, such as minimum wage laws and prohibitions on the sale of sexual favors, blood, body parts, and so on. To be sure, society as a whole may have an interest in such prohibitions, but our point is that individuals may favor them as devices for controlling their

---

7. For completeness, we can round out the set of contingencies by mentioning that either acceptance or rejection could be accompanied, not by regret, but by satisfaction: acceptance providing the satisfaction of, for example, extra money, or rejection providing that of upholding solidarity with the strikers. However, there is no challenge to the standard choice theory when an additional option yields additional benefits.

own behavior as well. Below, we offer a psychological hypothesis about why temptations always impose costs; here we emphasize that, insofar as a temptation is just an expansion of a choice set, it cannot reduce individual welfare within classical cost-benefit theory. If it is not just an expansion of a choice set, the problem for the normative theory is to explain what it is.

Note, also, that the best solution, in the utility-maximizing sense, would be to be compelled to accept a temptation, so that you can enjoy whatever makes it a temptation to begin with, but be absolved from the responsibility, which is to say, from the costs of violating a principle. ("The devil made me do it.") On the other side of the coin of reduced utility with an expanded choice set is the paradox of increasing utility when the choice set is reduced. ("Get thee behind me, Satan.")

If constraints on choice can improve welfare, there will be incentives for manufacturing constraints even though none "really" exist. A reformed alcoholic, for example, may have a policy of not walking down streets with many bars on them, but he is not really prevented in any physical way from doing so (Schelling, 1980, 1985). The constraint is a mental construction. In a similar vein, if a person can find an alternate principle that will permit a tempting action with no (or much reduced) cost, then he is in the best of all possible worlds: he may think he has really had his cake and eaten it too. When a reforming alcoholic accepts brandy after a dinner party, excusing himself by saying (and believing) that he did not want to offend his host, we may think he is kidding himself. This internal jockeying of choice sets makes no sense within rational choice theory, yet it is probably familiar to everyone.

Just as accepting or rejecting temptations has costs, costs may be incurred when someone is presented with an opportunity to *affirm* a principle. Consider two hypothetical situations:

Scenario 1: You buy an expensive bottle of wine that turns out to be mediocre; you drink it, making note not to buy it again.

Scenario 2: You order a comparable bottle in a restaurant (for the same price); after a great deal of thought, you decide not to send it back; however, your dinner has been spoiled.

The opportunity set in Scenario 2 is strictly larger, because it includes the option of returning the wine (not available in Scenario 1). How can this produce a lower "utility" level? By not sending the wine back, you have sacrificed a principle (something like "Get the fair value for purchases"), probably for pragmatic, cost-benefit reasons (e.g., avoiding a scene, embarrassment if the establishment refuses to take back the wine, personal acquaintance with the owner). The violation of principle exacts an additional cost, which is absent from Scenario 1.

When a rule is not followed, it must be because of cost-benefit reasons like these, because some other rule predominated, or because of physical restraint. It is our intuition that the subjective utility level will be *directly* related to how strong or compelling a reason there was for breaking the rule. If, for example, physical restraint was involved, rule breaking per se exacts little if any subjective cost. Or, to pick another example, homicide in self-defense is no crime at all, whereas homicide to avenge an uncollected debt is first-degree murder. The law mirrors our subjective sense of transgression. The subjective cost for breaking a rule rises with one's perceived margin for behaving otherwise. The perception of freedom of action is costly.

Costly Knowledge

A famous experiment of Stanley Milgram illustrates how information may hurt, in contradiction to the second theorem. In Milgram's experiment, many subjects discovered that they were willing to cause great apparent harm to another person, if pressed sufficiently hard by the experimenter, who represented himself as a scientist associated with Yale University.[8] For most subjects, participation in the experiment was a disagreeable experience, although they later felt it had taught them a useful lesson about themselves and other people (Milgram, 1974). Although the cost was partly due to the stress of the actual experiment, the more significant source of suffering was presumably the knowledge that one had failed a test of an important principle.

Some subjects obediently delivered the entire series of ostensibly painful and harmful shocks; other subjects abruptly resigned from the experiment. Two principles were evidently vying for control—something like "Do no harm," versus "Respect legitimate authority"—and subjects were guided by one or the other. It is relevant, though perhaps regrettable, that the defiant subjects were more unhappy about their experience as subjects than the obedient ones, as if to say that, on the av-

---

8. Subjects were instructed to give what appeared to them to be a series of progressively more severe electric shocks (but which were not really shocks at all) to another "subject" (an actor, in fact), who was seated in another room, but whose screams and pleadings could be clearly heard, as punishments for "incorrect" responses in a learning experiment.

erage, it costs more psychologically to violate the principle of respecting authority than of doing no harm. Subjects who behaved cruelly evidently felt less anguish than those who behaved admirably, but, in all cases, subjects suffered psychologically as they served in the study.

It is so easy to empathize with the discomfort of having participated in such an experiment that it takes some thought to see how puzzling it appears on strictly rational grounds. The procedure clearly provided fresh information about one's own behavioral dispositions, especially for those who complied with the instructions.[9] Although the information may be disturbing, it also provides an opportunity for the person to take morally corrective measures. A rational person may suffer as a result, but this is preferable, presumably, to remaining ignorant of one's profound susceptibility to authority. That is how many outsiders viewed Milgram's data, as a significant and welcome addition to our knowledge of human nature.

Curiously, it seems that the concerns over the ethical aspects of the study, of which there were many, derived from the fact that we all (i.e., the nonparticipants) benefited from its publication, at the expense of the actual participants (this issue is discussed in Herrnstein, 1974b). We are all now duly warned about following orders, etc., and, statistically, we know that there is a more than even chance that we would have given the entire sequence of shocks. But no one will lose sleep over this—indeed, one could even conjecture that no nonparticipant would feel great anguish even if *all* of the experimental subjects complied. There is a vast difference between knowing that one might have broken a principle, and actually doing so.

Valuable Misinformation

If information can be costly, it should come as no surprise that misinformation can be valuable, as illustrated by a hypothetical problem posed in Kahneman, Knetch, and Thaler (1986). Imagine yourself lying on the beach on a hot day, thinking about how nice it would be to have a cold beer. A friend offers to get one, from a place nearby—which, in the two versions of the story, is described as either a "fancy resort" or a "run-down grocery store." Not knowing the price of the beer, the friend requests instructions about the maximum amount that she should pay.

---

9. Milgram (1974) showed that virtually everyone thinks he would be defiant under experimental conditions that in fact elicit obedience from about two-thirds of the subjects.

Kahneman, Knetch, and Thaler report that the subjects' median reservation price was $1.50 for the grocery store, and $2.65 for the resort, the lower grocery-store price reflecting presumably a matter of principle (e.g., "Don't pay more than the customary price"). Suppose, now, that the friend goes to the grocery store and finds that the price is $1.75. What should she do? Purchase the beer, and explain, upon returning, that the grocery store was unfortunately closed, but that just down the road there was a fancy resort . . .

Multiple Selves and Ainslie's Theory of Rule Formation

It has been recognized for some time by a small but growing group of economic and behavioral theorists that in some situations, opportunities or information have negative value for the decision maker, even at the individual level. Aside from the work of Kahneman, Knetch, and Thaler (1986), which is concerned with the narrower issue of perceptions of fairness, these researchers have typically maintained the analytical techniques of the rational model, but viewed the person as being composed of several "subagents," each of which may have distinct preferences, as well as access to private information (Ainslie, 1975, 1982, 1986; Elster, 1986; Schelling, 1980; Thaler and Shefrin, 1981; Winston, 1980).

Observed from the outside, a single person would exhibit the behavior of a collective of rational agents, which, as formal game theory indicates, will not always be consistent with the one-person axioms of rationality. There are many games (the prisoner's dilemma being only the most famous example) in which individually rational play produces poor results and in which the elimination of some strategies would improve everyone's payoffs. The promise of this approach is to understand perceived suboptimalities in individual choice, by game-theoretic analogy, as the result of harmful strategic interactions between individually rational subsections of the self.

These attempts, however, also fall short, for lack of a key analytical instrument—something that would provide an analytics of *legal reasoning*. To develop this argument, we turn first to the work of George Ainslie, which, although less formal, contains an account of how rules and principles emerge from a strategic struggle for mastery within a single individual.

Ainslie proposes a theory of extraordinarily simple underlying assumptions. Because of an inherent characteristic of time perception—

documented experimentally (Ainslie, 1975)—people are prone to a systematic ambivalence about actions that create an initial benefit (or cost) followed, after some delay, by a larger cost (or benefit). When the moment of choice is relatively far away, we intend to give proper weight to the later consequence; however, when the moment of choice arrives, the smaller but earlier consequence overshadows the later one, causing an "impulsive" reversal of the original preference.

This intertemporal inconsistency sets the stage for an internal struggle between the individual's long- and short-term interests. Since the short-term interest is most often the "executive" (i.e., the one that actually makes the choice), it is naturally positioned to sabotage the intentions and plans of the long-term interest, unless the latter can alter the structure of rewards and punishments that are effective at the moment of choice. Ainslie has proposed a number of devices by which the long-term interest can prevail, but the most interesting, for our purposes, is the stratagem of *private rules,* or *side bets.*

The key step in the private rule strategy is to convince the short-term interest that the current decision will constitute a binding precedent for a long series of future choices. A current decision to, say, indulge some temptation becomes a sign that the temptation will be indulged on all such occasions. A single bite of the forbidden dessert, for instance, destroys the expectation that the diet will be followed in the future. The short-term interest is thus kept in line, as it were, by the threat of a severe and *immediate* loss in expectations for future outcomes.

Linking this account with our previous discussion, we could say that a principle is a mechanism serving the long-term interest, for the purpose of correcting the three types of cost-benefit mismatches discussed earlier. It amplifies the scale of the decision; it transforms unclear consequences into vivid ones; and, by staking future expectations on present choices, it brings the future to bear on the present.

But this argument now brings us to a central difficulty. Suppose that at the time of decision, the short-term interest is inclined to choose one way, but is reined in by the presence of some overarching rule. What is to prevent the short-term interest from interpreting the rule so that it does not apply to *this* particular case, which after all must differ in some respects from previous occasions when the rule was invoked, or, if that option is too farfetched, from invoking another, stronger principle that overrides the original one? At the moment of choice, the short-term interest is both the judge and the jury for any disputes over interpretation, which makes mysterious its presumed fidelity to the interpretations as they were originally laid down.

The Internal Marketplace of Principles

A missing element of the story, then, is some account of the psychological force by which a rule can fend off the challenge of immediate short-term interests. In addition, we need to understand the extent to which a rule can be stretched, or defeated by another dominating rule. These two issues are discussed in turn.

In Pavlovian, or classical, conditioning, the powers of one stimulus are, to some degree, imparted to another. In Pavlov's famous experiment on dogs, the power of dry meat powder to elicit salivation was transferred to a metronome or other arbitrary signal. The arbitrary signal is differentially associated or correlated with the natural (or previously conditioned arbitrary) stimulus.[10]

Much of the modern work on Pavlovian conditioning has dealt with the transference of motivational or emotion-eliciting powers across stimuli. For example, subjects will value an arbitrary signal that has been differentially associated with something they value to begin with. The ambiance of a room in which one has eaten numerous delicious meals acquires an aesthetic value in its own right. Collectors of books, documents, or old phonograph records treasure the feel, look, smell, and heft of their objects of interest, even though their fundamental value as collectibles depends on other attributes, such as rarity, age, and content.

Similarly, the arbitrary features of stimuli differentially associated with undesirable states of affairs become motivationally negative. The look and sound of a harsh teacher become disagreeable in their own right. We avoid his or her presence when we can. The disinfectant smell of hospitals has become a conditioned aversive stimulus for people who have endured fear, pain, or sorrow in that environment.

Society exploits our susceptibility to Pavlovian conditioning to inculcate social values. Social interactions are permeated with essentially ar-

---

10. Extensive research has refined Pavlov's rudimentary theory, according to which stimulus powers were transferred as a result of mere temporal contiguity and, in addition, the response being elicited was assumed to be unchanged in form. It is well known now that conditioned responses may differ systematically from unconditioned responses, and that mere contiguity is neither necessary nor sufficient to produce conditioning. An up-to-date account of this form of conditioning can be found in Mazur, 1986. For present purposes, the modern refinements can be set aside.

bitrary conventions of behavior and address, which arise in a history of conditioning that members of a community share. We are rewarded for politeness, punished for rudeness, and the differential association of those rewards and punishments with particular modes of behavior endow those modes of behavior with motivational power of their own. If we are successfully conditioned, we feel good about our good manners, bad about bad manners. That internalized increment or decrement may be sufficient to tip the scales toward politeness when the external circumstances might have evoked rudeness.

Conditioning attaches motivational power to arbitrary stimuli, and then, by some ill-understood process akin to reasoning or induction, that power may spread far from the particulars originally experienced, into a more or less logically coherent structure of stimuli and the perceived relations among them. Politeness becomes, not merely particular conditioned responses to particular stimuli, but a systematic approach to behavior in social settings. Having been conditioned aversively to avoid doing dangerous things like touching hot stoves or stepping off the curb into the street, we feel some of the attendant anxieties when we confront what we know or perceive to be new dangers. The value instilled in us by the external environment for good health habits may be elaborated internally into principles for staying fit.

On the face of it, the Pavlovian conditioning of values can be subsumed under standard economic analysis as a device for creating tastes. But, in fact, moral and prudential choices often involve deciding which of these articulated systems of internalized relations—or principles—applies in a given instance, then acting accordingly.

Our discussion of the selection among, or substitution of, alternate rules can be focused by means of the following example. Suppose that one particular Friday evening, a person chooses between going to an entertainment movie or a documentary film about famine in Africa (the proceeds of which will be donated to charity). It is the last showing for both movies. The person chooses the entertainment film, thus revealing his preference.

Now, to change the problem slightly, suppose that Friday after lunch, the person has the opportunity to take off early and see one of the two films (again, the last opportunity to see either one). He refuses the entertainment film, but might in fact go see the documentary. On narrow revealed preference grounds, this choice is irrational. Whatever the objective cost of taking the afternoon off, it is not affected by the movie that he sees: the benefits of either movie—hedonic, moral, or educational—

are no doubt technically separable in preference from the benefits that are derived from several hours' work.

It is not hard to reconstruct the internal argument that leads to the preference reversal. The person may have a rule against taking off early—especially for play—no doubt acquired from a history of conditioning. The documentary, however, is a hard case: In a sense it might be "work," especially if the film is not *too* enjoyable. In addition, there is a secondary principle at stake, namely, contributing to charitable causes, analogously acquired. Such ambiguities form the raw material for a decision to leave work, but they need to be shaped into a principle that has sufficient power to sanction the original desire to leave.

It is doubtful that rational calculation can shed much light on this process. Careful decision analysis may help us decide whether in a particular instance it is better to go against principle (one would rationally assess the costs of sacrificing the principle—a hard but not impossible task), but it does not help in deciding whether the principle actually applies to the issue at hand.

Decision analysis, which codifies the model of rational behavior and extends it into areas where uncertainty reigns, operates by resolving each problem into either a matter of preference or a matter of information. Let us start with the most obvious question. Does the documentary constitute work or play? This does not seem to be a matter of information, in the sense of uncertainty about some objective "state of the world." For example, if you think that it is work, and another person thinks that it is play, then your disagreement is not of the type that can be resolved by a scientific debate. There is no critical *fact* that establishes the correct labeling.

There is a temptation then to shift the question into the domain of tastes, which so often serves as the catch-all category for any nonobjective matter. But the thought processes one would go through in trying to classify the film do not seem to correspond to an interrogation of subjective *preference*. Quite possibly, the person has a good idea of how much he would enjoy each film; his doubt has to do with whether to allow himself this violation of principle.

What we see in the example, then, is a rather complete (and characteristic) fusion of tastes and beliefs. Our theory suggests that, up to a point, the probability of going to the documentary film may increase as the price (i.e., donation) rises. If one believes that seeing the film is work, then it can be seen (and enjoyed); if one believes that it is entertainment,

Table 15.1

	Action by decision analysis	Action by rule or principle
Goal	Maximize a given criterion	Discovery of, and adherence to, appropriate criterion
Method	Logical/empirical	Legal
Internal structure	Compensatory dimensions or attributes	Hierarchy of principles
Main cognitive activity	Assessment of multi-attribute value functions; prediction of uncertain events	Interpretation; judgments of grouping and similarity

then the enjoyment will be corrupted by the knowledge that a personal rule has been violated.

We also see that the arena in which principles compete is somewhat autonomous. If we were trying to convince someone to go see the film, we would not appeal to preferences, but would rather try to present him with an alternative interpretation or principle. This is the method of seduction through rhetoric: to provide arguments that sanction one's inclinations.

Divided-self models capture something important in describing moral ambivalence as a struggle—a game—between different, temporally seg-regated aspects of the individual. The problem is that they do not have much to say about the rules by which this game is conducted—and that is because the psychological processes that make the game possible are not well captured by variations on preferences and information (the twin building blocks of rational modeling).

In this chapter, we have tried to show that action by rules or princi-ples entails a distinct mode of decison making, irrespective of whether the rules pertain to moral or prudential concerns, whether the deci-sions are profound or trivial. Table 15.1 lists the main contrasts be-tween this legal/moral mode and the familiar procedures of decision analysis. Instead of evaluating trade-offs among the competing argu-ments of the utility function (such as risk versus expected return), the process revolves around a search for a unique principle that covers the decision at hand and that is not dominated by another more powerful principle.

This second mode of decision making has not yet benefited from the sustained economic and psychological research and attention that have been bestowed on the rational-agent model, perhaps because there is no normative theory to set the agenda. But if rules and principles do in fact arise because of fundamental defects in individual decision making, then a model of rule-governed action will have no shortage of applications in economics and the other social sciences.

References

Adby, P. R., and Dempster, M. A. H. (1974). *Introduction to optimization methods*. London: Chapman and Hall.

Ainslie, G. (1974). Impulse control in pigeons. *Journal of the Experimental Analysis of Behavior, 21,* 485–489.

Ainslie, G. (1975). Specious reward: A behavioral theory of impulsiveness and impulse control. *Psychological Bulletin, 82,* 463–496.

Ainslie, G. (1982). A behavioral economic approach to the defense mechanisms: Freud's energy theory revisited. *Social Science Information, 21,* 735–779.

Ainslie, G. (1986). Beyond microeconomics: Conflict among interests in a multiple self as a determinant of value. In J. Elster (ed.), *The multiple self.* Cambridge: Cambridge University Press.

Ainslie, G., and Herrnstein, R. J. (1981). Preference reversal and delayed reinforcement. *Animal Learning and Behavior, 9,* 476–482.

Allison, J. (1976). Contrast, induction, facilitation, suppression, and conservation. *Journal of the Experimental Analysis of Behavior, 25,* 185–198.

Allison, J., and Timberlake, W. (1974). Instrumental and contingent saccharin licking in rats: Response deprivation and reinforcement. *Learning and Motivation, 5,* 231–247.

Anderson, A. C. (1932). Time discrimination in the white rat. *Journal of Comparative Psychology, 13,* 27–55.

Arnold, V. I. (1973). *Ordinary differential equations.* Cambridge, Mass.: MIT Press.

Ashby, W. R. (1963). *An introduction to cybernetics.* New York: Wiley.

Ayala, F. J. (1972). Competition between species. *American Scientist, 60,* 348–357.

Ayala, F. J., and Campbell, C. A. (1974). Frequency-dependent selection. *Annual Review of Ecology and Systematics, 5,* 115–138.

Bailey, J. T., and Mazur, J. E. (1990). Choice behavior in transition: Development of preference for the higher probability of reinforcement. *Journal of the Experimental Analysis of Behavior, 53,* 409–422.

Baum, W. M. (1966). Choice in the rat as a function of probability and amount of reward. Doctoral dissertation, Harvard University.

Baum, W. M. (1974). On two types of deviation from the matching law: Bias and undermatching. *Journal of the Experimental Analysis of Behavior, 22,* 231–242.

Baum, W. M. (1979). Matching, undermatching, and overmatching in studies of choice. *Journal of the Experimental Analysis of Behavior, 32,* 269–281.

Baum, W. M., and Rachlin, H. C. (1969). Choice as time allocation. *Journal of the Experimental Analysis of Behavior, 12,* 861–874.

Baumol, W. J. (1977). *Economic theory and operations analysis.* Englewood Cliffs, N.J.: Prentice-Hall.

Becker, G. S. (1962). Irrational behavior and economic theory. *Journal of Political Economy, 70,* 1–13.

Becker, G. S. (1976). *The economic approach to human behavior.* Chicago: University of Chicago Press.

Becker, G. S., and Murphy, K. M. (1988). A theory of rational addiction. *Journal of Political Economy, 96,* 675–700.

Beier, E. M. (1988). Effects of trial-to-trial variation in magnitude of reward upon an instrumental running response. Doctoral dissertation, Yale University.

Bentham, J. (1789). *An introduction to the principles of morals and legislation.* London: D. Payne & Son.

Bitterman, M. E. (1965). Phyletic differences in learning. *American Psychologist, 20,* 396–410.

Blough, D. S. (1975). Steady state data and a quantitative model of operant generalization and discrimination. *Journal of Experimental Psychology: Animal Behavior Processes, 1,* 3–21.

Bohm, P. (1973). *Social efficiency: A concise introduction to welfare economics.* New York: Wiley.

Boland, L. A. (1981). On the futility of criticizing the neoclassical maximization hypothesis. *American Economic Review, 71,* 1031–1036.

Boole, G. (1854). *An investigation of the laws of thought, on which are founded the mathematical theories of logic and probabilities.* London: Walton and Maberly.

Booth, W. (1988). Social engineers confront AIDS. *Science, 242,* 1237–1238.

Bordley, R. F. (1983). A central principle of science: Optimization. *Behavioral Science, 28,* 53–64.

Bower, G. H., Fowler, H., and Trapold, M. A. (1959). Escape learning as a function of amount of shock reduction. *Journal of Experimental Psychology, 58,* 482–484.

Bradshaw, C. M., and Szabadi, E. (1988). Quantitative analysis of human operant behavior. In G. Davey and C. Cullen (eds.), *Human operant conditioning and behavior modification* (pp. 225–259). New York: Wiley.

Bradshaw, C. M., Szabadi, E., and Bevan, P. (1976). Behavior of humans in variable-interval schedules of reinforcement. *Journal of the Experimental Analysis of Behavior, 26,* 135–141.

Brody, J. E. (1988). Personal health. *New York Times,* August 25, p. 317.

Brown, R., and Herrnstein, R. J. (1975). *Psychology.* Boston: Little, Brown.

Bush, R. R., and Mosteller, F. (1955). *Stochastic models for learning.* New York: Wiley.

Campbell, B. A., and Kraeling, D. (1953). Response strength as a function of drive level and amount of drive reduction. *Journal of Experimental Psychology, 45,* 97–101.

Cannon, W. B. (1939). *The wisdom of the body* (rev. ed.). New York: Norton.

Caporael, L. R., Dawes, R. M., Orbell, J. M., and van de Kragt, A. J. C. (1989). Selfishness examined: Cooperation in the absence of egoistic incentives. *Behavioral and Brain Sciences, 12,* 683–739.

Catania, A. C. (1963a). Concurrent performances: A baseline for the study of reinforcement magnitude. *Journal of the Experimental Analysis of Behavior, 6,* 299–300.

Catania, A. C. (1963b). Concurrent performances: Reinforcement interaction and response independence. *Journal of the Experimental Analysis of Behavior, 6,* 253–263.

Catania, A. C. (1973). Self-inhibiting effects of reinforcement. *Journal of the Experimental Analysis of Behavior, 19,* 517–526.

Catania, A. C., and Reynolds, G. S. (1968). A quantitative analysis of the responding maintained by interval schedules of reinforcement. *Journal of the Experimental Analysis of Behavior, 111,* 327–383.

Christensen-Szalanski, J. J., Goldberg, A. D., Anderson, M. E., and Mitchell, T. R. (1980). Deprivation, delay of reinforcement, and the selection of behavioral strategies. *Animal Behaviour, 28,* 341–346.

Chung, S-H. (1965a). Effects of effort on response rate. *Journal of the Experimental Analysis of Behavior, 8,* 1–7.

Chung, S-H. (1965b). Effects of delayed reinforcement in a concurrent situation. *Journal of the Experimental Analysis of Behavior, 8,* 439–444.

Chung, S-H. (1966). Some quantitative laws of operant behavior. Doctoral dissertation, Harvard University.

Chung, S-H., and Herrnstein, R. J. (1967). Choice and delay of reinforcement. *Journal of the Experimental Analysis of Behavior, 10,* 67–74. Chapter 5 of the present book.

Clarke, B. C. (1979). The evolution of genetic diversity. *Proceedings of the Royal Society,* London, B, *205,* 453–474.

Commons, M. L. (1981). How reinforcement density is discriminated and scaled. In M. L. Commons and J. A. Nevin (eds.), *Quantitative analyses of behavior.* Vol. I. Cambridge, Mass.: Ballinger, pp. 51–85.

Commons, M. L., Herrnstein, R. J., and Rachlin, H. (eds.). (1982). *Quantitative analyses of behavior.* Vol. II: *Matching and maximizing accounts.* Cambridge, Mass.: Ballinger.

Commons, M. L., Woodford, M., and Ducheny, J. R. (1982). How reinforcers are aggregated in reinforcement-density discriminations and preference experiments. In M. L. Commons, R. J. Herrnstein, and H. Rachlin (eds.), *Quantitative analyses of behavior.* Vol. II: *Matching and maximizing accounts.* Cambridge, Mass.: Ballinger.

Conrad, D. G., and Sidman, M. (1956). Sucrose concentration as reinforcement for lever pressing by monkeys. *Psychological Reports, 2,* 381–384.

Cornsweet, T. N. (1970). *Visual perception*. New York: Academic Press.

Corso, J. F. (1967). *The experimental psychology of sensory behavior*. New York: Holt, Rinehart & Winston.

Crespi, L. P. (1942). Quantitative variation of incentive and performance in the white rat. *American Journal of Psychology, 55*, 467–517.

Davenport, J. W., Goodrich, K. P., and Hagguist, W. W. (1966). Effects of magnitude of reinforcement in *Macaca speciosa*. *Psychonomic Science, 4*, 187–188.

Davies, N. B., and Krebs, J. R. (1978). Introduction: Ecology, natural selection and social behavior. In J. R. Krebs and N. B. Davies (eds.), *Behavioural ecology: An evolutionary approach*. Sunderland, Mass.: Sinauer.

Davison, M., and McCarthy, D. (1988). *The matching law: A research review*. Hillsdale, N.J.: Erlbaum.

Dawes, R. M. (1988). *Rational choice in an uncertain world*. San Diego, Calif.: Harcourt Brace Jovanovich.

Dawkins, R. (1976). *The selfish gene*. Oxford: Oxford University Press.

Dawkins, R. (1980). Good strategy or evolutionarily stable strategy? In G. W. Barlow and J. Silverberg (eds.), *Sociobiology—beyond nature/nurture*. Boulder, Co.: Westview Press.

Dawkins, R. (1982). *The extended phenotype*. San Francisco: Freeman.

Dawkins, R., and Krebs, J. R. (1979). Arms races between and within species. *Proceedings of the Royal Society*. London, B, *205*, 489–511.

Deaton, A., and Muellbauer, J. (1980). *Economics and consumer behavior*. Cambridge: Cambridge University Press.

Deluty, M. Z. (1978). Self-control and impulsiveness involving aversive events. *Journal of Experimental Psychology: Animal Behavior Processes, 4*, 250–266.

de Villiers, P. A. (1974). The law of effect and avoidance: A quantitative relationship between response rate and shock-frequency reduction. *Journal of the Experimental Analysis of Behavior, 21*, 223–235.

de Villiers, P. A. (1977). Choice in concurrent schedules and a quantitative formulation of the law of effect. In W. K. Honig and J. E. R. Staddon (eds.), *Handbook of operant behavior* (pp. 233–287). Englewood Cliffs, N.J.: Prentice-Hall.

de Villiers, P. A., and Herrnstein, R. J. (1976). Toward a law of response strength. *Psychological Bulletin, 83*, 1131–1153. Chapter 2 of the present book.

Dews, P. B. (1960). Free-operant behavior under conditions of delayed reinforcement. I. CRF-type schedules. *Journal of the Experimental Analysis of Behavior, 3*, 221–234.

Di Lollo, V. (1964). Runway performance in relation to runway-goal-box similarity and changes in incentive amount. *Journal of Comparative and Physiological Psychology, 58*, 327–329.

Dinsmoor, J. A., and Hughes, L. H. (1956). Training rats to press a bar to turn off shock. *Journal of Comparative and Physiological Psychology, 49*, 235–238.

Donahoe, J. W. (1977). Some implications of a relational principle of reinforcement. *Journal of the Experimental Analysis of Behavior, 27*, 341–350.

Eisenberger, R., Karpman, M., and Trattner, J. (1967). What is the necessary and sufficient condition for reinforcement in the contingency situation? *Journal of Experimental Psychology, 74,* 342–350.

Elster, J. (1984). *Ulysses and the sirens.* Cambridge: Cambridge University Press.

Elster, J. (ed.). (1986). *The multiple self.* Cambridge: Cambridge University Press.

Estes, W. K. (1959). The statistical approach to learning theory. In S. Koch (ed.), *Psychology: A study of a science.* Vol. 2. New York: McGraw-Hill.

Etzioni, A. (1988). *The moral dimension: Toward a new economics.* New York: Free Press.

Fantino, E. (1966). Immediate reward followed by extinction vs. later reward without extinction. *Psychonomic Science, 6,* 233–234.

Faulhaber, G. R., and Baumol, W. J. (1988). Economists and innovators: Practical products of theoretical research. *Journal of Economic Literature, 26,* 577–600.

Ferster, C. B. (1953). Sustained behavior under delayed reinforcement. *Journal of Experimental Psychology, 45,* 218–224.

Ferster, C. B., and Skinner, B. F. (1957). *Schedules of reinforcement.* New York: Appleton-Century-Crofts.

Fisher, R. A. (1930). *The genetical theory of natural selection.* Oxford: Oxford University Press.

Fowler, H., and Trapold, M. A. (1962). Escape performance as a function of delay of reinforcement. *Journal of Experimental Psychology, 63,* 464–467.

Fraenkel, G. S., and Gunn, D. L. (1940). *The orientation of animals.* Oxford: Oxford University Press.

Frank, R. H. (1988). *Passions within reason: The strategic role of the emotions.* New York: Norton.

Friedman, M. (1989). An open letter to Bill Bennett. *Wall Street Journal,* September 7, p. A16.

Gallistel, C. R. (1969). The incentive of brain-stimulation reward. *Journal of Comparative and Physiological Psychology, 69,* 713–721.

Gentry, G. D., and Marr, M. J. (1980). Choice and reinforcement delay. *Journal of the Experimental Analysis of Behavior, 33,* 27–37.

Gibbon, J. (1977). Scalar expectancy theory and Weber's law in animal timing. *Psychological Review, 84,* 279–325.

Gilbert, R. M. (1972). Variation and selection of behavior. In R. M. Gilbert and J. R. Millenson (eds.), *Reinforcement: Behavioral analyses.* New York: Academic Press.

Goldstein, A., and Kalant, H. (1990). Drug policy: Striking the right balance. *Science, 249,* 1513–1521.

Goleman, D. (1989). Biology of brain may hold key for gamblers. *New York Times,* October 3, pp. C1, C11.

Gori, G. B. (1980). Less hazardous cigarettes: Theory and practice. In G. B. Gori and F. G. Bock (eds). *A safe cigarette?* Cold Spring Harbor, N.Y.: Cold Spring Harbor Laboratory, pp. 261–279.

Green, L., Kagel, J. H., and Battalio, R. C. (1982). Ratio schedules of reinforcement and their relation to economic theories of labor supply. In M. L. Commons, R. J. Herrnstein, and H. Rachlin (eds.), *Quantitative analyses of behavior*. Vol II: *Matching and maximizing accounts*. Cambridge, Mass.: Ballinger.

Green, L., and Snyderman, M. (1980). Choice between rewards differing in amount and delay: Toward a choice model of self-control. *Journal of the Experimental Analysis of Behavior, 34,* 135–148.

Gulliksen, H. (1934). A rational equation of the learning curve based on Thorndike's law of effect. *Journal of General Psychology, 11,* 395–434.

Guttman, N. (1953). Operant conditioning, extinction, and periodic reinforcement in relation to concentration of sucrose used as reinforcing agent. *Journal of Experimental Psychology, 46,* 213–223.

Guttman, N. (1954). Equal-reinforcement values for sucrose and glucose solutions compared with equal-sweetness values. *Journal of Comparative and Physiological Psychology, 47,* 358–361.

Hamilton, W. D. (1964). The genetical theory of social behavior, I and II. *Journal of Theoretical Biology, 7,* 1–16, 17–32.

Hardin, G. (1968). The tragedy of the commons. *Science, 162,* 1243–1248.

Harding, J., Allard, R. W., and Smeltzer, D. G. (1966). Population studies in predominately self-pollinated species, IX. Frequency-dependent selection in Phaseolus lunatus. *Proceedings of the National Academy of Sciences, 56,* 99–104.

Harrison, J. M., and Abelson, R. M. (1959). The maintenance of behavior by the termination and onset of intense noise. *Journal of the Experimental Analysis of Behavior, 2,* 23–42.

Heiner, R. A. (1983). The origin of predictable behavior. *American Economic Review, 73,* 560–595.

Helson, H. (1967). Perception. In H. Helson and W. Bevan (eds.), *Contemporary approaches to psychology*. Princeton, N.J.: Van Nostrand.

Herrnstein, R. J. (1958). Some factors influencing behavior in a two-response situation. *Transactions of the New York Academy of Science, 21,* 35–45.

Herrnstein, R. J. (1961). Relative and absolute strength of response as a function of frequency of reinforcement. *Journal of the Experimental Analysis of Behavior, 4,* 267–272. Chapter 1 of the present book.

Herrnstein, R. J. (1964). Aperiodicity as a factor in choice. *Journal of the Experimental Analysis of Behavior, 7,* 179–182.

Herrnstein, R. J. (1964). Will. *Proceedings of the American Philosophical Society, 108,* 455–458.

Herrnstein, R. J. (1969). Method and theory in the study of avoidance. *Psychological Review, 76,* 49–69.

Herrnstein, R. J. (1970). On the law of effect. *Journal of the Experimental Analysis of Behavior, 13,* 243–266.

Herrnstein, R. J. (1970). On the law of effect. *Journal of the Experimental Analysis of Behavior, 13,* 243–266.

Herrnstein, R. J. (1971). Quantitative hedonism. *Journal of Psychiatric Research, 8,* 399–412.

Herrnstein, R. J. (1974a). Formal properties of the matching law. *Journal of the Experimental Analysis of Behavior, 21,* 159–164.

Herrnstein, R. J. (1974b). Measuring evil. *Commentary*, June, pp. 82–88.

Herrnstein, R. J. (1977a). Doing what comes naturally. A reply to Professor Skinner. *American Psychologist, 32,* 1013–1016.

Herrnstein, R. J. (1977b). The evolution of behaviorism. *American Psychologist, 32,* 593–603.

Herrnstein, R. J. (1979). Derivatives of matching. *Psychological Review, 86,* 486–495. Chapter 3 of the present book.

Herrnstein, R. J. (1981). Self control as response strength. In C. M. Bradshaw, E. Szabadi, and C. F. Lowe (eds.), *Quantification of steady-state operant behaviour* (pp. 3–20). Amsterdam: Elsevier/North Holland Biomedical Press. Chapter 6 of the present book.

Herrnstein, R. J. (1982). Melioration as behavioral dynamism. In M. L. Commons, R. J. Herrnstein, and H. Rachlin (eds.), *Quantitative analyses of behavior.* Vol. II: *Matching and maximizing accounts.* Cambridge, Mass.: Ballinger. Chapter 4 of the present book.

Herrnstein, R. J. (1988). A behavioural alternative to utility maximization. In S. Maital (ed.), *Applied behavioral economics* (pp. 3–60). London: Wheatsheaf Books.

Herrnstein, R. J. (1989). Darwinism and behaviorism: Parallels and intersections. In A. Grafen (ed.), *Evolution and its influence* (pp. 35–61). London: Oxford University Press.

Herrnstein, R. J. (1990). Rational choice theory: Necessary but not sufficient. *American Psychologist, 45,* 356–367. Chapter 11 of the present book.

Herrnstein, R. J., and Heyman, G. M. (1979). Is matching compatible with reinforcement maximization on concurrent variable interval, variable ratio? *Journal of the Experimental Analysis of Behavior, 31,* 209–223.

Herrnstein, R. J., and Hineline, P. N. (1966). Negative reinforcement as shock-frequency reduction. *Journal of the Experimental Analysis of Behavior, 9,* 421–430.

Herrnstein, R. J., Loewenstein, G. F., Prelec, D., and Vaughan, W., Jr. (1993). Utility maximization and melioration: Internalities in individual choice. *Journal of Behavioral Decision Making, 6,* 149–185.

Herrnstein, R. J., and Loveland, D. H. (1972). Food avoidance in hungry pigeons, and other perplexities. *Journal of the Experimental Analysis of Behavior, 18,* 369–383.

Herrnstein, R. J., and Loveland, D. H. (1974). Hunger and contrast in a multiple schedule. *Journal of the Experimental Analysis of Behavior, 21,* 511–517.

Herrnstein, R. J., and Loveland, D. H. (1975). Maximizing and matching on concurrent ratio schedules. *Journal of the Experimental Analysis of Behavior, 24,* 107–116.

Herrnstein, R. J., and Mazur, J. E. (1987). Making up our minds: A new model of economic behavior. *The Sciences,* Nov./Dec., pp. 40–47.

Herrnstein, R. J., and Prelec, D. (1991). Melioration: A theory of distributed choice. *Journal of Economic Perspectives, 5,* 137–156. Chapter 14 of the present book.

Herrnstein, R. J., and Prelec, D. (1992a). A theory of addiction. In G. F. Loewenstein and J. Elster (eds.), *Choice over time.* New York: Russell Sage Publications. Chapter 9 of the present book.

Herrnstein, R. J., and Prelec, D. (1992b). Melioration. In G. F. Loewenstein and J. Elster (eds.), *Choice over time*. New York: Russell Sage Publications.

Herrnstein, R. J., Prelec, D., and Vaughan, W., Jr. (1986). An intra-personal prisoners' dilemma. Paper presented at the IX Symposium on Quantitative Analysis of Behavior: Behavioral Economics, Harvard University.

Herrnstein, R. J., and Vaughan, W., Jr. (1980). Melioration and behavioral allocation. In J. E. R. Staddon (ed.), *Limits to action: The allocation of individual behavior*. New York: Academic Press.

Herrnstein, R. J., and Vaughan, W., Jr. (1984). Evolutionary and behavioral stability. *Behavioral and Brain Sciences, 7,* 10.

Heyman, G. M. (1979). A Markov model description of changeover probabilities on concurrent variable-interval schedules. *Journal of the Experimental Analysis of Behavior, 31,* 41–51.

Heyman, G. M. (1982). Is time collection unconditional behavior? In M. L. Commons, R. J. Herrnstein, and H. Rachlin (eds.), *Quantitative analyses of behavior*. Vol. II: *Matching and maximizing accounts*. Cambridge, Mass.: Ballinger.

Heyman, G. M., and Herrnstein, R. J. (1986). More on concurrent interval-ratio schedules: A replication and review. *Journal of the Experimental Analysis of Behavior, 46,* 331–351.

Heyman, G. M., and Luce, R. D. (1979a). Operant matching is not a logical consequence of maximizing reinforcement rate. *Animal Learning and Behavior, 7,* 133–140.

Heyman, G. M., and Luce, R. D. (1979b). Reply to Rachlin. *Animal Learning and Behavior, 7,* 269–270.

Hineline, P. N. (1977). Negative reinforcement and avoidance. In W. K. Honig and J. E. R. Staddon (eds.), *Handbook of operant behavior*. Englewood Cliffs, N.J.: Prentice-Hall.

Hirsh, M. W., and Smale, S. (1974). *Differential equations, dynamical systems, and linear algebra*. New York: Academic Press.

Hirshleifer, J. (1982). Evolutionary models in economics and law: Cooperation versus conflict strategies. *Research in law and economics, 4,* 1–60.

Hirshleifer, J. (1985). The expanding domain of economics. *American Economic Review, 75,* 53–68.

Homans, G. C. (1974). *Social behavior: Its elementary forms* (rev. ed.). New York: Harcourt Brace Jovanovich.

Houston, A. I. (1986). The matching law applies to wagtails foraging in the wild. *Journal of the Experimental Analysis of Behavior, 45,* 15–18.

Houston, A. I., and McNamara, J. M. (1988). A framework for the functional analysis of behaviour. *Behavioral and Brain Sciences, 11,* 117–163.

Hull, C. L. (1943). *Principles of behavior*. New York: Appleton-Century.

Hull, D. L. (1973). *Darwin and his critics*. Cambridge, Mass.: Harvard University Press.

Hume, D. (1826). *The philosophical works* (Vol. 2). Edinburgh: Black and Tait. (Original work published 1777.)

Hursh, S. R. (1978). The economics of daily consumption controlling food- and water-reinforced responding. *Journal of the Experimental Analysis of Behavior, 29,* 475–491.

Hutt, P. J. (1954). Rate of bar pressing as a function of quality and quantity of food reward. *Journal of Comparative and Physiological Psychology, 47*, 235–239.

Irwin, S. (1990). *Drugs of abuse.* Tempe, Ariz.: Do It Now Foundation.

Janet, A. (1895). Considerations mechaniques sur l'evolution et le problem des especes. *C.R. 3me Congr. Int. Zool.,* Leyde, 136–145.

Janis, I., and Mann, L. (1977). *Decision making: A psychological analysis of conflict, choice, and commitment.* New York: Free Press.

Jellinek, E. M. (1960). *The disease concept of alcoholism.* New Haven, Conn.: College and University Press.

Jevons, W. S. (1871). *The theory of political economy.* London: Macmillan.

Kahneman, D., Knetch, J., and Thaler, R. (1986). Fairness as a constraint on profit seeking: Entitlements in the market. *American Economic Review, 76*, 728–741.

Kahneman, D., Slovic, P., and Tversky, A. (eds.). (1982). *Judgment under uncertainty: Heuristics and biases.* Cambridge: Cambridge University Press.

Keesey, R. E. (1962). The relation between pulse frequency, intensity, and duration and the rate of responding for intracranial stimulation. *Journal of Comparative and Physiological Psychology, 55*, 671–678.

Keesey, R. E. (1964). Duration of stimulation and the reward properties of hypothalamic stimulation. *Journal of Comparative and Physiological Psychology, 58*, 201–207.

Keesey, R. E., and Kling, J. W. (1961). Amount of reinforcement and free-operant responding. *Journal of the Experimental Analysis of Behavior, 4*, 125–132.

Killeen, P. R. (1972). The matching law. *Journal of the Experimental Analysis of Behavior, 17*, 489–495.

Killeen, P. R. (1981). Averaging theory. In C. M. Bradshaw, E. Szabadi, and C. F. Lowe (eds.), *Quantification of steady-state operant behaviour.* Amsterdam: Elsevier.

Killeen, P. R. (1982). Incentive theory. In *Nebraska symposium on motivation, 1981.* Lincoln: University of Nebraska Press.

Kline, M. (1972). *Mathematical thought from ancient to modern times.* New York: Oxford University Press.

Koestler, A. (1968). *The ghost in the machine.* New York: Macmillan.

Koob, G. F., and Bloom, F. E. (1988). Cellular and molecular mechanisms of drug dependence. *Science, 242*, 715–723.

Kraeling, D. (1961). Analysis of amount of reward as a variable in learning. *Journal of Comparative and Physiological Psychology, 54*, 560–565.

Krebs, J. R. (1978). Optimal foraging: Decision rules for predators. In J. R. Krebs and N. B. Davies (eds.), *Behavioural ecology: An evolutionary approach.* Sunderland, Mass.: Sinauer.

Krebs, J. R., and Davies, N. B. (eds.) (1978). *Behavioral ecology: An evolutionary approach.* Sunderland, Mass.: Sinauer.

Kudadjie-Gyanfi, E., and H. Rachlin (1996). Temporal patterning in choice among delayed outcomes. *Organizational Behavior and Human Decision Processes, 65*, 61–67.

Lakatos, I. (1970). Falsification and the methodology of scientific research programmes. In I. Lakatos and A. Musgrave (eds.), *Criticism and the growth of knowledge.* Cambridge: Cambridge University Press.

Leeming, F. C., and Robinson, J. E. (1973). Escape behavior as a function of delay of negative reinforcement. *Psychological Reports, 32,* 63–70.

Leonard, D. (1989). Market behavior of rational addicts. *Journal of Economic Psychology, 10,* 117–144.

Lesieur, H. R. (1989). Current research into pathological gambling and gaps in the literature. In H. J. Shaffer, S. A. Stein, B. Gambino, and T. N. Cummings (eds.), *Compulsive gambling: Theory, research, and practice.* Lexington, Mass.: Lexington Books, pp. 225–248.

Lipsey, R. G., and Steiner, P. O. (1972). *Economics* (3rd ed.). New York: Harper & Row.

Logan, F. A. (1960). *Incentive.* New Haven: Yale University Press.

Logan, R. A., and Spanier, D. (1970). Relative effect of delay of food and water reward. *Journal of Comparative Physiology and Psychology, 72,* 102–104.

Logue, A. W. (1988). Research on self-control: An integrating framework. *Behavioral and Brain Sciences, 11,* 665–709.

Luce, R. D. (1959). *Individual choice behavior.* New York: Wiley.

Luce, R. D. (1988). Rank-dependent, subjective expected-utility representations. *Journal of Risk and Uncertainty, 1,* 305–332.

Luce, R. D. (1989). Linear utility models with rank- and sign-dependent weights. Manuscript.

Luce, R. D. (1990). Rational versus plausible accounting equivalences in preference judgment. *Psychological Science, 1,* 225–234.

Machina, M. J. (1987). Decision-making in the presence of risk. *Science, 236,* 537–543.

Mackintosh, N. J. (1974). *The psychology of animal learning.* New York: Academic Press.

Mackintosh, N. J. (1975). A theory of attention: Variations in the associability of stimuli with reinforcement. *Psychological Review, 82,* 276–298.

Mansfield, E. (1975). *Microeconomics* (2nd ed.). New York: Norton.

Mansfield, R. J. W. (1976). Visual adaptation: Retinal transduction, brightness and sensitivity. *Vision Research, 16,* 679–690.

Margolis, H. (1987). *Patterns, thinking, and cognition: A theory of judgment.* Chicago: University of Chicago Press.

Marks, L. E. (1974). *Sensory processes: The new psychophysics.* New York: Academic Press, 1974.

May, R. M. (1983). Review of *Food Webs* by S. L. Pimms. *Science, 220,* 295–296.

Maynard Smith, J. (1974). The theory of games and the evolution of animal conflicts. *Journal of Theoretical Biology, 47,* 209–221.

Maynard Smith, J. (1976). Evolution and the theory of games. *American Scientist, 64,* 41–45.

Maynard Smith, J. (1978). Optimization theory in evolution. *Annual Review of Ecological Systems, 9,* 31–56.

Maynard Smith, J. (1982). *Evolution and the Theory of Games.* Cambridge: Cambridge University Press.

Maynard Smith, J., and Price, G. R. (1973). The logic of animal conflict. *Nature, 246,* 15–18.

Mazur, J. E. (1975). The matching law and quantifications related to Premack's principle. *Journal of Experimental Psychology: Animal Behavior Processes, 1,* 374–386.

Mazur, J. E. (1977). Quantitative studies of reinforcement relativity. *Journal of the Experimental Analysis of Behavior, 25,* 137–149.

Mazur, J. E. (1981). Optimization theory fails to predict performance of pigeons in a two-response situation. *Science, 214,* 823–825.

Mazur, J. E. (1984). Tests of an equivalence rule for fixed and variable reinforcer delays. *Journal of Experimental Psychology: Animal Behavior Processes, 10,* 426–436.

Mazur, J. E. (1985). Probability and delay of reinforcement as factors in discrete-trial choice. *Journal of the Experimental Analysis of Behavior, 43,* 341–351.

Mazur, J. E. (1986a). Choice between single and multiple delayed reinforcers. *Journal of the Experimental Analysis of Behavior, 46,* 67–77.

Mazur, J. E. (1986b). *Learning and behavior.* Englewood Cliffs, N.J.: Prentice-Hall.

Mazur, J. E. (1987). An adjusting procedure for studying delayed reinforcement. In M. L. Commons, J. E. Mazur, J. A. Nevin, and H. Rachlin (eds.), *Quantitative analyses of behavior.* Vol. 5: *The effect of delay and of intervening events on reinforcement value* (pp. 55–73). Hillsdale, N.J.: Erlbaum.

Mazur, J. E., and Hastie, R. (1978). Learning as accumulation: A reexamination of the learning curve. *Psychological Bulletin, 85,* 1256–1275.

Mazur, J. E., and Herrnstein, R. J. (1988). On the functions relating delay, reinforcer value, and behavior. *Behavioral and Brain Sciences, 11,* 690–691. Chapter 7 of the present book.

Mazur, J. E., and Logue, A. W. (1978). Choice in a "self-control" paradigm: Effects of a fading procedure. *Journal of the Experimental Analysis of Behavior, 30,* 11–17.

Mazur, J. E., Snyderman, M., and Coe, D. (1985). Influences of delay and rate of reinforcement on discrete-trial choice. *Journal of Experimental Psychology: Animal Behavior Processes, 11,* 565–575.

Mazur, J. E., Stellar, J. R., and Waraczynski, M. (1987). Self-control choice with electrical stimulation of the brain as a reinforcer. *Behavioural Processes, 15,* 143–153.

Mazur, J. E., and Vaughan, W., Jr. (1987). Molar optimization versus delayed reinforcement as explanations of choice between fixed-ratio and progressive-ratio schedules. *Journal of the Experimental Analysis of Behavior, 48,* 251–261.

McCoy, J. W. (1979). The origin of the "adaptive landscape" concept. *American Naturalist, 113,* 610–613.

McDowell, J. J., and Kessel, R. (1979). A multivariate rate equation for variable-interval performance. *Journal of the Experimental Analysis of Behavior, 31,* 267–283.

McSweeney, F. K. (1978). Prediction of concurrent key-peck treadle-press responding from simple schedule performance. *Animal Learning and Behavior, 6,* 444–450.

Milgram, S. (1974). *Obedience to authority: An experimental view.* New York: Harper & Row.

Miller, H. L., Jr. (1976). Matching-based hedonic scaling in the pigeon. *Journal of the Experimental Analysis of Behavior, 26,* 335–347.

Moffat, G. H., and Koch, D. L. (1973). Escape performance as a function of delay of reinforcement and inescapable U.S. trials. *Psychological Reports, 32,* 1255–1261.

Moskowitz, H. R. (1970). Ratio scales of sugar sweetness. *Perception & Psychophysics, 7,* 315–320.

Murray, C. (1988). *In pursuit of happiness and good government.* New York: Simon & Schuster.

Myers, D. L., and Myers, L. E. (1977). Undermatching: A reappraisal of performance on concurrent variable-interval schedules of reinforcement. *Journal of the Experimental Analysis of Behavior, 25,* 203–214.

Myerson, J., and Miezin, F. M. (1980). The kinetics of choice: An operant systems analysis. *Psychological Review, 87,* 160–174.

Nadelmann, E. A. (1989). Drug prohibition in the United States: Costs, consequences, and alternatives. *Science, 245,* 939–947.

Navarick, D. J., and Fantino, E. (1976). Self-control and general models of choice. *Journal of Experimental Psychology: Animal Behavior Processes, 2,* 75–87.

Nelson, R. R., and Winter, S. G. (1982). *An evolutionary theory of economic change.* Cambridge, Mass.: Harvard University Press.

Nevin, J. A. (1969). Interval reinforcement of choice behavior in discrete trials. *Journal of the Experimental Analysis of Behavior, 12,* 875–885.

Nevin, J. A. (1979). Overall matching *versus* momentary maximizing: Nevin (1969) revisited. *Journal of Experimental Psychology: Animal Behavior Processes, 5,* 300–306.

Nevin, J. A. (1979). Reinforcement schedules and response strength. In M. D. Zeiler and P. Harzem (eds.), *Reinforcement and the organization of behavior.* New York: Wiley.

Orford, J. (1985). *Excessive appetites: A psychological view of addictions.* Chichester, UK: Wiley and Sons.

Osborne, S. R. (1977). The free food (contrafreeloading) phenomenon: A review and analysis. *Animal Learning and Behavior, 5,* 221–235.

Passell, P. (1989). Why it pays to be generous. *New York Times,* January 25, p. D2.

Pattison, E. M., Sobell, M. B., and Sobell, L. B. (1977). *Emerging concepts of alcohol dependence.* New York: Springer.

Perin, C. T. (1943). A quantitative investigation of the delay-of-reinforcement gradient. *Journal of Experimental Psychology, 32,* 37–51.

Pierce, C. H., Hangord, P. V., and Zimmerman, J. (1972). Effects of different delay of reinforcement procedures on variable-interval responding. *Journal of the Experimental Analysis of Behavior, 18,* 141–146.

Pitt, D. E. (1988). Judge hears addict's plea to help him. *New York Times,* December 23, pp. B1–B2.

Pollack, R. A. (1970). Habit formation and dynamic demand functions. *Journal of Political Economy, 78,* 745–763.

Prelec, D. (1982). Matching, maximizing, and the hyperbolic reinforcement feedback function. *Psychological Review, 89,* 189–230.

Prelec, D. (1983a). Choice and behavior allocation. Doctoral dissertation. Harvard University.

Prelec, D. (1983b). The empirical claims of maximization theory: A reply to Rachlin, and Kagel, Battalio, and Green. *Psychological Review, 90,* 385–389.

Prelec, D. (1989). Decreasing impatience: Definition and consequences. Harvard Business School Working Paper 90-015.

Prelec, D., and Herrnstein, R. J. (1991). Preferences or principles: Alternative guidelines for choice. In R. J. Zeckhauser (ed.), *Strategy and choice.* Cambridge, Mass.: MIT Press, pp. 319–340. Chapter 15 of the present book.

Prelec, D., and Herrnstein, R. J. (1992). Melioration. In G. F. Loewenstein and J. Elster (eds.), *Choice over time.* New York: Russell Sage.

Premack, D. (1959). Toward empirical behavior laws: I. Positive reinforcement. *Psychological Review, 66,* 219–233.

Premack, D. (1965). Reinforcement theory. In D. Levine (ed.), *Nebraska symposium on motivation* (vol. 13). Lincoln: University of Nebraska Press.

Premack, D. (1971). Catching up with common sense or two sides of a generalization: Reinforcement and punishment. In R. Glaser (ed.), *The nature of reinforcement.* New York: Academic Press.

Pyke, G. H., Pulliam, H. R., and Charnov, E. L. (1977). Optimal foraging: A selective review of theory and tests. *Quarterly Review of Biology, 52,* 137–154.

Rachlin, H. (1970). *Introduction to modern behaviorism.* San Francisco: W. H. Freeman.

Rachlin, H. C. (1971). On the tautology of the matching law. *Journal of the Experimental Analysis of Behavior, 15,* 249–251.

Rachlin, H. (1980). Economics and behavioral psychology. In J. E. R. Staddon (ed.), *Limits to action.* New York: Academic Press.

Rachlin, H. (in press). The value of temporal patterns in behavior. *Current Directions.*

Rachlin, H., and Green, L. (1972). Commitment, choice, and self-control. *Journal of the Experimental Analysis of Behavior, 17,* 15–22.

Rachlin, H., Battalio, R., Kagel, J., and Green, L. (1981). Maximization theory in behavioral psychology. *Behavioral and Brain Sciences, 4,* 371–417.

Rachlin, H., Green, L., Kagel, J. H., and Battalio, R. C. (1976). Economic demand theory and psychological studies of choice. In G. H. Bower (ed.), *The psychology of learning and motivation* (vol. 10). New York: Academic Press.

Rachlin, H., Logue, A. W., Gibbon, J., and Frankel, M. (1986). Cognition and behavior in studies of choice. *Psychological Review, 93,* 33–45.

Rescorla, R. A., and Wagner, A. R. (1972). A theory of Pavlovian conditioning: Variations in the effectiveness of reinforcement and nonreinforcement. In A. H. Black and W. F. Prokasy (eds.), *Classical conditioning. II: Current research and theory.* New York: Appleton-Century-Crofts.

Restle, F., and Greeno, J. G. (1970). *Introduction to mathematical psychology.* Reading, Mass.: Addison-Wesley.

Reynolds, G. S. (1961). Relativity of response rate and reinforcement frequency in a multiple schedule. *Journal of the Experimental Analysis of Behavior, 2,* 179–184.

Reynolds, G. S. (1963). On some determinants of choice in pigeons. *Journal of the Experimental Analysis of Behavior, 6,* 53–59.

Richards, R. W., and Hittesdorf, W. M. (1978). Inhibitory stimulus control under conditions of signalled and unsignalled delay of reinforcement. *Psychological Record, 28,* 615–625.

Rosen, R. (1967). *Optimality principles in biology.* New York: Plenum.

Roy, A., Adinoff, B., Roehrich, L., Lamparski, D., Custer, R., Lorenz, V., Barbaccia, M., Guidotti, A., Costa, E., and Linnoila, M. (1988). Pathological gambling: A psychobiological study. *Archives of General Psychiatry, 45,* 369–373.

Roy, A., De Long, J., and Linnoila, M. (1985). Extraversion in pathological gamblers: Correlates with indexes of noradrenergic function. *Archives of General Psychiatry, 46,* 679–681.

Rozin, P. (1977). The significance of learning mechanisms in food selections: Some biology, psychology, and sociology of science. In L. M. Barker, M. R. Best, and M. Domjan (eds.), *Learning mechanisms in food selection.* Waco, Tex.: Baylor University Press, pp. 557–589.

Rozin, P. (1982). Human food selection: The interaction of biology, culture, and individual experience. In L. M. Barker (ed.), *The psychobiology of human food selection.* Bridgeport, Conn.: AVI, pp. 225–254.

Rozin, P., and Fallon, A. E. (1981). The acquisition of likes and dislikes for foods. In J. Solms and R. L. Hall (eds.), *The role of food components in food acceptance.* Zurich: Forster, pp. 35–48.

Rozin, P., and Schiller, D. (1980). The nature and acquisition of a preference for chili peppers. *Motivation and Emotion, 4,* 77–101.

Russell, M. A. H. (1979). Tobacco dependence: Is nicotine rewarding or aversive? In N. A. Krasnegor (ed.), *Cigarette smoking as a dependence process.* Rockville, Md.: National Institute of Drug Abuse, pp. 100–122.

Samuelson, P. A. (1947). *Foundations of economic analysis.* Cambridge, Mass.: Harvard University Press.

Schachter, S. (1971). Some extraordinary facts about obese humans and rats. *American Psychologist, 26,* 129–144.

Schelling, T. C. (1978). *Micromotives and macrobehavior.* New York: Norton.

Schelling, T. C. (1980). The intimate contest for self-command. *Public Interest, 60,* 94–118.

Schelling, T. C. (1985). Enforcing rules on oneself. *Journal of Law, Economics, and Organization, 1,* 357–374.

Schneider, B. A. (1969). A two-state analysis of fixed-interval responding in the pigeon. *Journal of the Experimental Analysis of Behavior, 12,* 677–687.

Schoemaker, P. J. H. (1982). The expected utility model: Its variants, purposes, evidence and limitations. *Journal of Economic Literature, 20,* 529–563.

Schoener, T. W. (1969). Optimal size and specialization in constant and fluctuating environments: An energy-time approach. *Diversity and stability in ecological systems.* Brookhaven Symposia in Biology: No. 22.

Schoener, T. W. (1971). Theory of feeding strategies. *Annual Review of Ecology and Systematics, 2,* 369–404.

Schrier, A. M. (1963). Sucrose concentration and response rates of monkeys. *Psychological Reports, 12,* 666.

Schrier, A. M. (1965). Response rates of monkeys under varying conditions of sucrose reinforcement. *Journal of Comparative and Physiological Psychology, 59,* 378–384.

Scitovsky, T. (1975). *The joyless economy: An inquiry into human satisfaction and consumer dissatisfaction.* New York: Oxford University Press.

Seward, J. P., Shea, R. A., Uyeda, A. A., and Raskin, D. C. (1960). Shock strength, shock reduction, and running speed. *Journal of Experimental Psychology, 60,* 250–254.

Shaffer, H. J., Stein, S. A., Gambino, B., and Cummings, T. N. (eds.). *Compulsive gambling: Theory, research, and practice.* Lexington, Mass.: Lexington Books.

Shimp, C. P. (1966). Probabilistically reinforced choice behavior in pigeons. *Journal of the Experimental Analysis of Behavior, 9,* 443–455.

Shimp, C. P. (1969). Optimal behavior in free-operant experiments. *Psychological Review, 76,* 97–112.

Shimp, C. P. (1975). Perspectives on the behavioral unit: Choice behavior in animals. In W. K. Estes (ed.), *Handbook of learning and cognitive processes.* Vol. 2: *Conditioning and behavior theory.* Hillsdale, N.J.: Erlbaum.

Shull, R. L., Spear, D. J., and Bryson, A. E. (1981). Delay or rate of food delivery as a determiner of response rate. *Journal of the Experimental Analysis of Behavior, 35,* 129–143.

Sidman, M. (1953). Avoidance conditioning with brief shock and no exteroceptive warning signal. *Science, 118,* 157–158.

Sidman, M. (1966). Avoidance behavior. In W. K. Honig (ed.), *Operant behavior: Areas of research and application.* New York: Appleton-Century-Crofts.

Silberberg, A., Hamilton, B., Ziriax, J. M., and Casey, J. (1978). The structure of choice. *Journal of Experimental Psychology: Animal Behavior Processes, 4,* 368–398.

Silver, M. P., and Pierce, C. H. (1969). Contingent and noncontingent response rates as a function of delay of reinforcement. *Psychonomic Science, 14,* 231–232.

Simon, H. A. (1955). A behavioral model of rational choice. *Quarterly Journal of Economics, 59,* 99–118.

Simon, H. A. (1956). Rational choice and the structure of the environment. *Psychological Review, 63,* 129–138.

Simon, H. A. (1957). *Models of man, social and rational: Mathematical essays on rational human behavior.* New York: Wiley.

Sizemore, O. J., and Lattal, K. A. (1978). *Journal of the Experimental Analysis of Behavior, 30,* 169–175.

Skinner, B. F. (1935). The generic nature of the concepts of stimulus and response. *Journal of General Psychology, 12,* 40–65.

Skinner, B. F. (1938). *The behavior of organisms.* New York: Appleton Century Co.

Skinner, B. F. (1981). Selection by consequences. *Science, 213,* 501–504.

Slatkin, M. (1978). On the equilibrium of fitnesses by natural selection. *American Naturalist, 112,* 845–859.

Smale, S. (1980). *The mathematics of time.* New York: Springer-Verlag.

Smith, A. (1776). *Inquiry into the nature and causes of the wealth of nations.* London: W. Strahan & T. Cadell.

Solnick, J. V., Kannenberg, C. H., Eckerman, D. A., and Waller, M. B. (1980). *Learning and Motivation, 11,* 61–77.

Solomon, R. L. (1977). An opponent-process theory of motivation. IV: The affective demands of drug addiction. In J. D. Mazer and M. E. P. Seligman (eds.), *Psychopathology: Laboratory models.* San Francisco: W. H. Freeman.

Solomon, R. L., and Corbit, J. D. (1973). An opponent-process theory of motivation. II: Cigarette addiction. *Journal of Abnormal Psychology, 81,* 158–171.

Solomon, R. L., and Corbit, J. D. (1974). An opponent-process theory of motivation: I. Temporal dynamics of affect. *Psychological Review, 81,* 119–145.

Staddon, J. E. R. (1972). Temporal control and the theory of reinforcement schedules. In R. M. Gilbert and J. R. Millenson (eds.), *Reinforcement: Behavioral analyses.* New York and London: Academic Press.

Staddon, J. E. R. (1977). On Herrnstein's equation and related forms. *Journal of the Experimental Analysis of Behavior, 28,* 163–170.

Staddon, J. E. R., and Motheral, S. (1978). On matching and maximizing in operant choice experiments. *Psychological Review, 85,* 436–444.

Staddon, J. E. R., and Simmelhag, V. L. (1971). The "superstition" experiment: A reexamination of its implications for the principles of adaptive behavior. *Psychological Review, 78,* 3–43.

Sternberg, S. (1963). Stochastic learning theory. In R. D. Luce, R. R. Bush, and E. Galanter (eds.), *Handbook of mathematical psychology.* Vol. 2. New York: Wiley.

Stigler, G. J. (1966). *The theory of price* (3rd ed.). New York: Macmillan.

Stigler, G. J., and Becker, G. S. (1977). De gustibus non est disputandum. *American Economic Review, 67,* 76–90.

Strotz, R. H. (1956). Myopia and inconsistency in dynamic utility maximization. *Journal of Economic Studies, 23,* 166–180.

Stubbs, D. A., and Pliskoff, S. S. (1969). Concurrent responding with fixed relative rate of reinforcement. *Journal of the Experimental Analysis of Behavior, 12,* 887–895.

Tarpy, R. M. (1969). Reinforcement difference limen (RDL) for delay in shock escape. *Journal of Experimental Psychology, 79,* 116–121.

Tarpy, R. M., and Koster, E. D. (1970). Stimulus facilitation of delayed-reward learning in the rat. *Journal of Comparative and Physiological Psychology, 71,* 147–151.

Thaler, R. (1980). Toward a positive theory of consumer choice. *Journal of Economic Behavior and Organization, 1,* 39–60.

Thaler, R., and Shefrin, H. M. (1981). An economic theory of self-control. *Journal of Political Economy, 89,* 392–410.

Thorndike, L. L. (1898). Animal intelligence: An experimental study of the associative processes in animals. *Psychological Review Monograph Supplements 11,* no. 4 (whole, no. 8).

Thurstone, L. L. (1919). The learning curve equation. *Psychological Monographs, 26* (3, whole no. 114).

Thurstone, L. L. (1930). The learning function. *Journal of General Psychology, 3,* 469–493.

Timberlake, W., and Allison, J. (1974). Response deprivation: An empirical approach to instrumental performance. *Psychological Review, 81,* 146–164.

Tolman, E. C. (1938). The determiners of behavior at a choice point. *Psychological Review, 45,* 1–41.

Tversky, A., and Kahneman, D. (1981). Framing of decisions and the psychology of choice. *Science, 211,* 453–458.

Tversky, A., and Kahneman, D. (1986). Rational choice and the framing of decisions. *The Journal of Business, 59,* S251–S278.

Vaughan, W., Jr. (1976). Optimization and reinforcement. Doctoral dissertation. Harvard University.

Vaughan, W., Jr. (1981). Melioration, matching, and maximization. *Journal of the Experimental Analysis of Behavior, 36,* 141–149.

Vaughan, W., Jr. (1982). Choice and the Rescorla-Wagner model. In M. L. Commons, R. J. Herrnstein, and H. Rachlin (eds.), *Quantitative analyses of behavior.* Vol. II: *Matching and maximizing accounts.* Cambridge, Mass.: Ballinger.

Vaughan, W., Jr. (1984). Giving up the ghost. *Behavioral and Brain Sciences, 7,* 501.

Vaughan, W., Jr. (1985). Choice: A local analysis. *Journal of the Experimental Analysis of Behavior, 43,* 383–405.

Vaughan, W., Jr. (1986). Choice and punishment: A local analysis. In M. L. Commons, J. E. Mazur, J. A. Nevin, and H. Rachlin (eds.), *Quantitative analyses of behavior.* Vol. 5: *Effects of delay and intervening events on reinforcement value.* Hillsdale, N.J.: Erlbaum.

Vaughan, W., Jr., and Herrnstein, R. J. (1987). Stability, melioration, and natural selection. In L. Green and J. H. Kagel (eds.), *Advances in behavioral economics* (vol. 1, pp. 185–215). Norwood, N.J.: Ablex. Chapter 10 of the present book.

Vaughan, W., Jr., Kardish, T. A., and Wilson, M. (1982). Correlation versus contiguity in choice. *Behavioral Analysis Letters, 2,* 153–160.

Vaughan, W., Jr., and Miller, H. L., Jr. (1984). Optimization versus response-strength accounts of behavior. *Journal of the Experimental Analysis of Behavior, 42,* 337–348.

Vuchinich, R. E., and Tucker, J. A. (1988). Contributions from behavioral theories of choice to an analysis of alcohol abuse. *Journal of Abnormal Psychology, 97*, 181–195.

Walsh, G. R. (1975). *Methods of optimization.* London: John Wiley.

Watson, J. B. (1917). The effect of delayed feeding upon learning. *Psychobiology, 1*, 51–59.

Weizsäcker, C. C., von (1971). Notes on endogenous change of tastes. *Journal of Economic Theory, 3*, 345–372.

Wells, L. (1988). Conspicuous consumers. *New York Times Sunday Magazine,* November 20, p. 90.

Williams, B. A. (1976). The effects of unsignalled delayed reinforcement. *Journal of the Experimental Analysis of Behavior, 26*, 441–449.

Williams, B. A. (1988). Reinforcement, choice, and response strength. In R. C. Atkinson, R. J. Herrnstein, G. Lindzey, and R. D. Luce (eds.), *Stevens' handbook of experimental psychology.* Vol. 2 (pp. 167–244). New York: Wiley.

Williams, B. A., and Fantino, E. (1978). Effects on choice of reinforcement delay and conditioned reinforcement. *Journal of the Experimental Analysis of Behavior, 29*, 77–86.

Williams, G. C. (1966). *Adaptation and natural selection.* Princeton, N.J.: Princeton University Press.

Wilson, D. S. (1980). *The natural selection of populations and communities.* Menlo Park, Calif.: Benjamin/Cummings.

Wilson, J. Q., and Herrnstein, R. J. (1985). *Crime and human nature.* New York: Simon & Schuster.

Winston, G. C. (1980). Addiction and backsliding. *Journal of Economic Behavior and Organization, 1*, 295–324.

Woods, P. J., Davidson, E. H., and Peters, R. J. (1964). Instrumental escape conditioning in a water tank: Effects of variation in drive stimulus intensity and reinforcement magnitude. *Journal of Comparative and Physiological Psychology, 57*, 466–470.

Woods, P. J., and Holland, C. H. (1966). Instrumental escape conditioning in a water tank: Effects of constant reinforcement at different levels of drive stimulus intensity. *Journal of Comparative and Physiological Psychology, 62*, 403–408.

Wright, S. (1932). The roles of mutation, inbreeding, crossbreeding and selection in evolution. In D. F. Jones (ed.), *Proceedings of the Sixth International Congress of Genetics* (Vol. 1). Menasha, Wisc.: Brooklyn (N.Y.) Botanic Garden.

Wynne-Edwards, V. C. (1962). *Animal dispersion in relation to social behaviour.* Edinburgh: Oliver and Boyd.

Zeaman, D. (1949). Response latency as a function of the amount of reinforcement. *Journal of Experimental Psychology, 39*, 466–483.

Zeckhauser, R. (1986). Comments: Behavioral versus rational economics: What you see is what you conquer. *Journal of Business, 59*, S435–S449.

Zuriff, G. E. (1985). *Behaviorism: A conceptual reconstruction.* New York: Columbia University Press.

Index